The Compleat Plattmaker

PUBLISHED UNDER THE AUSPICES OF THE
WILLIAM ANDREWS CLARK MEMORIAL LIBRARY
UNIVERSITY OF CALIFORNIA, LOS ANGELES

The Compleat Plattmaker

Essays on Chart, Map, and Globe Making
in England in the Seventeenth and
Eighteenth Centuries

Edited by
NORMAN J. W. THROWER
Clark Library Professor, 1972–1973

1978
UNIVERSITY OF CALIFORNIA PRESS
BERKELEY • LOS ANGELES • LONDON

University of California Press
Berkeley and Los Angeles, California
University of California Press, Ltd.
London, England

To Franklin D. Murphy, M.D., Chancellor of the University of California, Los Angeles, 1960–1968, who breathed new life into the Clark Library, as he did to all parts of the UCLA campus.

CONTENTS

PREFACE

The William Andrews Clark Memorial Library of the University of California, Los Angeles, is dedicated, particularly, to the study of British culture in the seventeenth and eighteenth centuries. Within these geographical, thematic, and temporal limits, the period 1640 to 1750 and English literature receive special attention. This emphasis appropriately reflects the strength of the Clark Library holdings in the field. Other subjects that were interests of the Clark family, represented by their collections and largesse, include law, music, art, and science.

William Andrews Clark, Junior, founded the library as a tribute to his father Senator William Andrews Clark, the copper king, whose remarkable house at Butte, Montana, has recently been opened to the public. The Clark Library now honors both father and son. The story of its transfer, with a substantial endowment, to the embryonic Los Angeles campus of the University of California in 1926 and of its subsequent growth can be read in the various published reports of the Clark Library. This growth has paralleled that of the UCLA campus that lies some ten miles westward and oceanward of the Clark Library.

The unrivaled Dryden and other literary collections at the Clark Library are emblematic of the love of its founder for belles lettres. William Andrews Clark, Junior, was a lawyer by profession and gave his alma mater, the University of Virginia, its beautiful Clark Hall, formerly the Law School but now used as an environmental center. A consuming passion of the younger

Clark, as of other members of the Clark family, was music. He owned and played Thomas Jefferson's violin, founded and generously supported the Los Angeles Philharmonic Orchestra, and built an elegant chamber music room at the Clark Library. The princely gifts of the Clark family to the Corcoran Gallery in Washington, D.C., and the Clark Library building itself, as well as its oil paintings and murals and collections of prints and fine printing, all testify to the founder's taste in the visual arts. The Clark interest in science is less well known; it focused especially upon what Joseph Needham has called the sciences of the heavens and the earth—cosmology or cosmography.

In the splendid marble entrance hall of the Clark Library the founder's interests are expressed by a series of six paintings by Allyn Cox, the well-known muralist. These include, among other symbols, a terrestrial globe and an armillary sphere. These orbs, which occupy central and opposite positions in the series of vignettes, represent respectively, geoscience and astronomy. The concern of the Clarks with earth science is typified by the rich collection of minerals given by the family to the state of Montana. As for astronomy, Mr. Clark built an observatory on his estate adjacent to the library and employed his own resident astronomer. Some years ago the observatory building was razed because it was unsafe. Recently, however, through the kindness of Peggy Christian, a friend of the UCLA libraries, an early illustrated brochure on the observatory was added to the collection of Clark ephemera. This brochure reads in part:

The Clark Observatory, W. A. Clark Jr., 2205 W. Adams St., Los Angeles, California.

Cards of admission are freely granted and the services of an instructor provided to those who apply in person at the downtown office of the Curator . . . Mars F. Baumgardt.

Interesting Features

Telescope . . . Brashear refractor of exquisite workmanship with six-inch objective.
Radium Maps of the principal constellations and the Milky Way.
Model of the Solar System.
Meteor Collection, principally from Canyon Diablo, Arizona. One of the meteors weighs 357 pounds.
Globular model of the Moon, 40 inches in Diameter.

Accordingly, it is appropriate that scientific subjects in addition to literature, law, music, and the visual arts should be included as part of the Clark Library program. Indeed, facets of early science, as well as other aspects of culture, have been discussed in several Clark Library Saturday Seminars, as perusal of the growing list of these events will reveal. Also, significantly, a portrait of Sir Isaac Newton graces the paneled walls of the North Reading Room at the Clark Library, while John Dryden presides over the South Reading Room. Further, the writings of the Honourable Robert Boyle, among other seventeenth-century spirits of the Scientific Revolution, are well represented among Clark materials. Although Mr. Clark's observatory no longer exists, its large telescope is now in the Department of Astronomy, and the collection of meteorites at the Institute of Geophysics at UCLA.

As a result of increased support from UCLA and its chancellors, the programs of the Clark Library have been considerably expanded in recent years. Thus the Saturday Seminars have grown from one (or less) per year in the 1950s to an average of about four per year at the present time. Similarly the Summer Seminar is now a regular program in which half-a-dozen young postdoctoral fellows spend six weeks under the tutelage of a senior scholar. Also the award of a yearlong Clark Graduate Fellowship enables at least one advanced student from UCLA to work at the Clark on a doctoral dissertation. In addition, the Senior Research Fellowship brings one or two scholars to the Clark Library for several months each year. These activities have been in progress for some time, but recently have moved at a quickened pace or have been considerably enlarged.

An entirely new feature was added to the program when the chancellor at UCLA instituted the Clark Library Professorship in 1968. The intention is for the position to be held by a professor of English for one year, alternating annually with a professor representing some other discipline appropriate to the Clark Library. Among his duties the Clark Professor is expected to hold a series of talks, in effect a continuing seminar. Funds are available so that speakers from any part of the country and even from overseas may be invited to address the seminar. Usually the talks are given in a series, the audiences consisting of scholars and graduate students from UCLA and various southern California institutions, with a sprinkling drawn from as far away as the San

Francisco Bay Area. Typically the talks are held on Friday afternoons and a visitor may give a single presentation, or several over a period of a few weeks. The Clark Library Professor is encouraged to assemble and edit a series of essays arising from the Friday or Saturday Seminars or other special Clark Library occasions.

The present volume contains six essays that resulted from various Clark Library activities. Two are by speakers of the Friday Seminars that were held when the present editor was a Clark Library Professor and three were Saturday Seminars that he arranged and moderated. The final essay is based on a Seminar paper that he delivered in 1968; it is the only one of the six previously published under Clark Library auspices, but it is now out of print.

A high point of the academic year in which the history of cartography was especially emphasized at the Clark Library, was the two-week visit of Dr. Helen Wallis. Her present post is one that has been held by a succession of eminent map librarians going back to the establishment of an independent Department of Maps in the British Museum in 1867. Dr. Wallis, a trained geographer and historian, has specialized in the history of cartography and globe making, and in the history of European exploration and settlement. In a happy combination of her interests, she spoke on Saturday morning, 10 February 1973, on the subject of "English Globes and Geography in the Days of Pepys and Swift." The large and distinguished audience included, appropriately, the late Dr. William Matthews and Mrs. Matthews. William Matthews, Professor Emeritus of English at UCLA, was the principal editor (with Robert Latham, librarian of the Pepys Library at Cambridge University) and Lois Matthews was the text assistant of the California edition of *Pepys's Diary*. Helen Wallis in her essay for the present volume has emphasized the earlier period—that of Samuel Pepys who, as the energetic secretary of the Royal Navy and fellow and president of the Royal Society, was much concerned with matters relating to charting and mapping. His observations on these subjects were as perspicacious as his remarks on many other aspects of culture in seventeenth-century England. Directly or indirectly the influence of Pepys provides a continuing thread throughout these essays.

On Saturday, 3 March 1972, Miss Jeanette Black and Dr. Thomas R. Smith gave talks at the Clark Library in a seminar under the general title, "Some Aspects of Seventeenth-Century

Cartography." Such a seminar at the Clark Library has been
charmingly described by Professor Miner in the preface to the
series of essays that he edited, and so the format need not be
detailed here.

In his essay, Professor Smith provides us with a very insightful
look at the final phase, in England, of the portolan chart trade.
This cartographic genre had its shadowy beginnings in Italy c.
1300, after the development of (and presumably related to) the
magnetic compass in Amalfi, Pisa, and Genoa. The early portolan
chart trade also flourished in the Balearic Islands under the patronage of the kings of Aragon and was transmitted thence to
Portugal where it underwent significant change. In its altered
form the portolan chart both facilitated and served as a record
of the geographical discoveries of the Iberians after 1415. The
Thames School, discussed by Professor Smith, represents a continuation of this tradition from the late sixteenth to the early
eighteenth century with the English chartmakers attempting
through their manuscript delineations, to compete with the more
successful Dutch printed sea charts of this period. Eventually in
England, as elsewhere, the manuscript sea chart gave way to the
printed form and some of the later Thamesmen aided in this
transition. Professor Smith has not only established the professional relationship between the various practitioners of the
Thames School in his essay but regales us with a picaresque frag-
ment that sheds light on the personal life of a member of the guild
and his family.

It is to the author of the third essay, Jeannette Black, as Professor Smith informs us, that credit is due for the appellation
"Thames School," a term approved by the late R. A. Skelton.
In her own essay, Miss Black is more concerned with the English
surveyors and mapmakers who went overseas than those who
stayed at home and compiled, engraved, and published the original material supplied to them from the field. The English were
very much dependent on the Dutch for maps of Europe, including even those of the coasts of Britain in this period, as was more
than once remarked upon and deplored. However, there was
frequently more originality in mapping endeavors in the colonies
than in the homeland. This has often been the case in the history
of cartography, witness the innovative British cadastral and thematic mapping in Ireland, or their epic topographic mapping
accomplishments in India. It appears that in remote stations, distant from centralized control, engineers and others were equal to

the great opportunities afforded them in many fields including mapping. Sometimes they produced manuscript materials that only became well known through compilation, reproduction, and publication on the Continent where engraving techniques were better developed at this time. Later on, some surveyors, like James Cook, who began his professional charting career in what is now eastern Canada, were to spend part of the year overseas in mapping and part at home working up the results of the previous field season. It was only after this development that British mapmakers really outdistanced the continental cartographers in general mapping, as they had earlier in some aspects of thematic mapping.

Dr. David A. Woodward and Professor Coolie Verner were senior research fellows at the Clark Library during the spring of 1973. They each gave four lectures in a series on seventeenth- and eighteenth-century cartography, the former emphasizing surveying and printing techniques, and the latter publishing and cartobibliography. Coolie Verner has worked extensively on the history of cartography in the Clark period, and in his May 1973 presentations, drew upon this long acquaintance with the field. In his essay in the present volume Professor Verner focuses on the important figure of John Seller. However, as Professor Verner informs us, perhaps Seller's contributions may be overrated in comparison with those of some of his contemporaries, especially John Thornton. Seller was in a position to have his name attached to many maps and atlases, to which others may have claim to an important share of the credit. This situation is not unknown today but the cartobibliography of modern maps must be different from that of the seventeenth and eighteenth centuries because, as Richard Gardiner has suggested, of the anonymity of many modern cartographers, especially those employed by commercial houses and government departments. Institutional or company names only are printed on many, perhaps most, contemporary maps, globes, and atlases.

David Woodward has been very close to the topics he presented at his March and April Seminars at the Clark Library which are summarized and condensed in his essay for this volume. Both Coolie Verner and David Woodward, as senior research fellows, were able to make particularly good use of the UCLA Special Collections and Clark Library map resources. These include the newly acquired Clark Library Collection of seventeenth- and early eighteenth-century distance maps by John Adams and his successors which, among many other materials, Dr. Woodward discusses and illustrates in his essay. He also treats the technical matters of engraving and surveying, giving credence, when considered with the other essays in the volume, to our main title —*The Compleat Plattmaker*. In this title, in place of "Plattmaker," the alliterative term *Cartographer* was avoided. Although, as Cornelis Koeman has told us recently, the term *cartography* was apparently coined as early as 1557, popularization of this general designation for a maker of maps followed the inspired moment when the second Visconde de Santarem in a letter from Paris to Lisbon reinvented the word (*cartographia*) in 1839. As John Wolter informs us, "the word 'Cartographie' was first printed in the *Bulletin de la Société de Geographie de Paris*, in December 1840 . . . and first appears in English in the *Journal of the Royal Geographical Society* for 1843."

Any consideration of English mapping in the seventeenth and eighteenth centuries which pretends to completeness must take into account the beginnings of thematic mapping and pioneer work in this field of Edmond Halley. Curiously, though, Halley's signal contributions to cartography have been neglected in the past by historians of cartography, including that indefatigable researcher, E. G. R. Taylor, who was more interested in the "lesser men" than in the great scientists and philosophers. The final essay, on Halley, was presented at the morning session of a Clark Library Saturday Seminar held on 27 April 1968. In the afternoon, Professor Clarence J. Glacken of the University of California, Berkeley, gave a talk entitled "On Chateaubriand's Journey from Paris to Jerusalem, 1806–07" and the two essays were published under the title, *The Terraqueous Globe*. In the present reprinting of my earlier paper there are some minor revisions, resulting from further research on Halley. At a time when only little progress was being made in cartography in England by others, Halley developed mapping techniques that were later adapted by such scientists as Alexander von Humboldt and have since been universally adopted. As well as being one of the greatest of English scientists, Halley was a cartographic innovator of the very first rank. Happily for the editor, Halley's dates (1656–1742) are bracketed by the Clark Library period.

In the prefatory remarks to his Clark Library presentation, Professor Smith acknowledged the support of Franklin D. Murphy, formerly chancellor of UCLA, and of Robert Vosper, formerly university librarian at UCLA, the continuing director of the Clark Library and now professor of Library and Information Science at UCLA in these words.

Library collections of this sort stand as monuments to the collabor-
ative meshings of interests, abilities and resources of numerous individuals. There is the collector—usually a bookman, sometimes a
benefactor. The librarian, in addition to being an efficient and dis-
criminating conservator, should be a perceptive sensor, even stim-
ulator of the scholarly activities within his community. The dealer
is the purveyor of good things which excite the mind and stimulate
the senses. These individuals need not know each other and, in anti-
quarian matters, often do not even share the same century. But it is
the active and personal collaboration of builders and patrons of
collections that bring the excitement and satisfaction of development.
In this context and in these days of structured university systems,
public funds and private endowments, the university administrator
can do much to stimulate interest and thus supplement the amenities
available to the librarian and scholar. The annals of American librar-
ianship record several instances of particularly fruitful cooperation
between librarians and their chancellors, and a particularly notable
example has characterized this university. But my own university
was ahead of UCLA in this particular, because it was at the University of Kansas in the 1950s that the felicitous collaboration of Robert
Vosper as Librarian and Franklin Murphy as Chancellor first began.
There, as here, the university, its library and its scholars benefited
and momentum was established which has continued in these more
difficult times.

   I would like to add my deep appreciation to these scholar-
administrators and to their successors. They and other members
of the Clark Library Committee undergirded the scholarly endeavors represented by this volume and those of my predecessors
in the Clark Library Professorship, aptly called "a celestial experience" by Professor Swedenberg, the first incumbent. I would
also like to thank William E. Conway, Librarian, and Edna C.
Davis, formerly Associate Librarian, and other members of the
friendly and always helpful staff of the Clark Library. The
authors of the first five essays who came as visitors to the Clark
Library deserve special thanks for their presentations, which are
now available to a larger audience in published form. All of us
are indebted to William Andrews Clark, Junior, who lamented
that his professional and business affairs took him too often away
from the things he really loved but who, through his gifts, made
it possible for others to work in and enjoy the cultivated at-
mosphere which he created.

                                        N. J. W. T.

The Clark Library
1 March 1975

CONTRIBUTORS

Jeannette D. Black, retired, formerly Curator of Maps at the John Carter Brown Library, Providence, Rhode Island, U.S.A.

Thomas R. Smith, Professor and formerly Chairman of the Department of Geography-Meteorology at the University of Kansas, Lawrence, Kansas, U.S.A.

Norman J. W. Thrower, Professor of Geography at the University of California, Los Angeles, California, U.S.A.

Coolie Verner, Professor in the Adult Education Research Centre, University of British Columbia, Vancouver, Canada.

Helen M. Wallis, Map Librarian, the Map Library of the British Library (formerly Superintendent of the Map Room of the British Museum), London, England.

David A. Woodward, Curator of Maps and Program Director of the Hermon Dunlap Smith Center for the History of Cartography, The Newberry Library, Chicago, Illinois, U.S.A.

I

GEOGRAPHIE IS BETTER THAN DIVINITIE. MAPS, GLOBES, AND GEOGRAPHY IN THE DAYS OF SAMUEL PEPYS

Helen M. Wallis

On 13 June 1660 the mathematician William Oughtred, then eighty-six years old, cried out for joy at the news of the Restoration of the Monarchy and the accession of King Charles II: " 'And are yee sure he is restored? Then give me a glasse of Sack to drinke his Sacred Majestie's health.' His spirits were then quite upon the wing to fly away." He died with the toast of the Restoration on his lips (as John Aubrey records), severing with his death the last link with the Elizabethan mathematicians.[1] Many such toasts must have been drunk in the course of that summer. Men like Samuel Pepys who had grown up with Puritan leanings and connections, shared the general relief that the rigors of the Interregnum were over.

The conditions of the previous twenty years had not favored the development of the geographical arts and sciences. In the days of King Charles I when the court at Whitehall was celebrated for its patronage of the arts, there had been no comparable royal sponsorship of mathematics, science, or cosmography. During the Interregnum printing had almost ceased. Map making and globe making had been in decline for many years, and by the

---

[1] The story was told by Oughtred's great friend, the mathematical instrument maker Ralph Greatorex, who "conceived [of Oughtred] he dyed with joy for the comeing-in of the King, which was the 29th of May before." Oliver Lawson Dick, ed., *Aubrey's Brief Lives* (London, 1950), p. 255. The comment on the Elizabethan link is E. G. R. Taylor's, *The Mathematical Practitioners of Tudor and Stuart England 1485–1714* (Cambridge, 1954), p. 99.

1640s the neglect of mathematics had become a matter of public concern. "Alasse! what a sad case it is that in this great and opulent kingdome there is no publick encouragement for the excelling in any Profession but that of the Law and Divinity," exclaimed Dr. Sanderson, the Lord Bishop of Lincoln, to the Cambridge mathematician Dr. John Pell, c. 1640.[2] Pepys, educated at St. Paul's School and Magdalene College, Cambridge, felt it necessary in July 1662, when nearly thirty years old, and clerk of the acts, to take arithmetic lessons from a Mr. Cooper, mate of the *Royall Charles* ("my first attempt being to learn the Multiplication table.")[3] In geography too he found his education defective. When a contract for the Forest of Dean had to be negotiated with Sir John Winter, secretary to the Queen Mother and principal entrepreneur in the Forest, Pepys, "turned to the forrest of Deane in Speedes mapps; and there he [Sir John Winter] showed me how it lies . . . and many other things worth my knowing; and I do perceive that I am very short in my business by not knowing many times the geographicall part of my business."[4] John Speed's *Theatre of the Empire of Great Britaine* first published in 1611[–1612], and the earliest published atlas of the British Isles, ran through a series of editions up to 1676. Pepys acquired the edition of 1625, bound up with the companion world atlas, *A Prospect of the most famous Parts of the World* (1631).

Geographical studies were of three kinds: first, the use of the globes, terrestrial and celestial, a subject mainly mathematical in treatment; secondly, cosmography and geography, whose descriptive literature covered a wide range of topics; and collections of travels. In the second class, Peter Heylyn's *Cosmographie*, published in 1652, remained by far the most popular work for fifty years or more, running to some six editions before 1700, a revised "second" edition appearing in 1703. The *Cosmographie* provided a model of seventeenth-century descriptive geography. Of the anecdotes that enlivened it, one of the best known is Heylyn's account of "a pretty accident" that befell him in January 1640 [i.e., 1641], at a time when his royalist and religious activities as a high churchman and divine had brought him into trouble with the House of Commons and had won him enemies.

[2] O. L. Dick, *Aubrey's Brief Lives*, p. 230.
[3] 4 July 1662. Robert Latham and William Matthews, eds., *The Diary of Samuel Pepys*, III (Berkeley and Los Angeles, 1970), 131.
[4] 20 June 1662. Ibid., pp. 114–115.

"Amongst others, I was then incountred in my passage from Westminster to Whitehall, by a tall big Gentleman, who thrusting me rudely from the wall, and looking over his shoulder on me in a scornfull manner, said in a hoarse voyce these words, '*Geographie is better than Divinitie*,' and so passed along. Whether his meaning were, that I was a better *Geographer* than *Divine*; or that *Geographie* had been a Study of more credit and advantage to me in the eyes of men, than *Divinitie* was like to prove, I am not able to determine."[5] The incident had its influence on Heylyn, "in drawing me to look back on those younger studies, in which I was resolved to have dealt no more." He was thus inspired to enlarge his *Microcosmus, or a Little Description of the Great World* (first published in 1621) into the *Cosmographie*. Among collections of travels, Hakluyt's *Principal Navigations* (1598–1600) and *Purchas his Pilgrimes* (1625) remained the standard authorities; but *Purchas* was now out of favor, as Pepys records: "*Purchas* his work was sold for 4 or 5s. before the Fire of London, being valued but as so much waste paper. Remember Mr. Evelyn's rectifying Bab. May's low opinion of his book by obliging him to read one volume of it."[6]

The Restoration ushered in a period of intellectual and economic progress reminiscent of the Elizabethan age. The arts and sciences benefited from the establishment of stable government and from the rapid expansion of economic activity. A prosperous landed gentry was now buying books, prints, and pictures. The scholars and craftsmen returning home from exile on the continent had been stimulated by their contact with new ideas, such as the philosophical and scientific speculations of Descartes. Deploring their dependence on foreign works, Englishmen found, moreover, that the atlases of the rival Dutch firms of Jansson and Blaeu were no longer readily available for purchase. This offered an opportunity for ingenious and industrious citizens to establish their reputations and make a livelihood by meeting the needs of an expanding middle class. The Fire of London in 1666 destroyed many private libraries and accentuated this demand. Although it also destroyed the stock of a number of dealers and practitioners such as that of John Ogilby (who was left with a capital of £5),[7]

5 Peter Heylyn, *Cosmographie in foure Bookes* . . . (London, 1652), sig. A.3ᵛ.

6 J. R. Tanner, ed., *Samuel Pepys's Naval Minutes*, Publications of the Navy Records Society 60 (London, 1926). 123. Baptist May, a friend of John Evelyn's, was keeper of the Privy Purse to King Charles II.

7 O. L. Dick, *Aubrey's Brief Lives*, p. 221. *Calendar of State Papers, Domestic Series*, . . . *1666–1667* (London, 1864), Sept. 1666, pp. 171–172, items 109 and 110.

these men found an even brisker business open to them once they had set up again.

One of the returned emigrés from the Netherlands was Joseph Moxon (1627–1700). He had spent the years of his youth from the age of ten to sixteen in the Dutch cities of Delft and Rotterdam, where his father James Moxon had settled as a printer. Returning to London in 1646, they set up in partnership as printers. Around 1650 Joseph established himself as a globe and instrument maker and mapseller "at the signe of the Atlas in Cornhill." Visiting Amsterdam in 1652 to study printing techniques, Moxon brought back W. J. Blaeu's handbook to globes, then just published, and translated it into English as *A Tutor to Astronomy and Geography, or an Easie and Speedy Way to understand the use of both Globes*, which he printed and published in 1654, selling it at his shop, "where you may also have globes of all sizes."[8] This work, now very rare, was superseded when Moxon published, in 1659, a totally new handbook, *A Tutor to Astronomie and Geographie, or an Easie and speedy way to know the Use of both the Globes, Celestial and Terrestrial, in six books*. Specially designed for an English as opposed to a Dutch public, it was Moxon's most popular book on globes, and by 1698 had run to five editions, with another so-called "fifth" edition appearing also in 1699.

By 1673 Moxon was advertising a range of globes, from "Globes, 26 Inches diameter. The price 20 l. the pair," to "Concave Hemispheres of the Starry Orb, which serves for a Case to a Terrestrial Globe of 3 Inches diameter, made portable for the Pocket. Price 15 sh."; and "Spheres, according to the Copernican Hypothesis, both General and Particular, 20 Inches Diameter. Price of both to l."; and "Spheres, according to the Ptolomaick Systeme, near 15 Inches Diameter. Price 3 l., . . . 8 Inches Diameter price 1 l. 10 sh."[9] For the sale of his globes Moxon made

[8] Carey S. Bliss, *Some Aspects of Seventeenth Century English Printing with Special Reference to Joseph Moxon* (Los Angeles, 1965), p. 16. Bliss's suggestion that Moxon *may* have printed the work is correct, for Moxon in his address "To the Reader" in *A Tutor to Astronomie* (London, 1659) refers to the earlier work, which "I formerly printed," and explains why he thought it necessary to write a new textbook for an English public. *A Tutor . . .* (1654), as described by Bliss, differs also from Moxon's, *A Tutor to Astronomy & Geography, or the Use of the Copernican Spheres; in two Books* (London, 1665), which also was based on Blaeu's book on globes, with adaptations to suit an English readership. A copy is in the British Library (532.f.18).

[9] A catalog of "Globes Cælestial and Terrestrial," made and sold by Moxon, 1657 (B.L. 529.g.24), sig. Q 1ᵛ. The globes are also advertised in a catalog

use of the popular device of the lottery. His complete range of globes was advertised in his "Proposals for a New Mathematical Lottery, without Hazard: wherein will be no Blanks, but manifest Advantages to the Adventurers. There will be delivered out 1000 Tickets at Twenty Shillings per Ticket...." It was proposed that "the First and Last drawn ... shall have each, besides the Goods drawn against the Number, a Benefit of 40 l. in such Globes, and other Things, best made up, as are set forth in the above-written Scheme." The lottery was to be drawn on Tuesday, 27 September next, and Dr. Hooke of Gresham College (curator at the Royal Society) and Mr. Samuel Newton, master of the Mathematical School in Christ's Hospital, were among those named as supervisors. Tickets were to be had at "J. Moxon's at the Atlas in Warwick-lane, at Man's Coffee-House at Charing-Cross, Abel Roper Book-seller, at the Black-Boy in Fleet-street, and at Lloyd's Coffee-House in Lombard-street."[10] On another occasion Moxon arranged for the raffle of his pair of 24-inch globes, "being the largest and best that have been done in England, the Frames being twisted Banisters standing on 12 Pedestals coloured in Oil of a Marble Red, the Globes curiously pasted and coloured, and the Stars gilt, and varnished with the whitest hard Varnish, the Brasen Meridian, Hour-Circle, Indexes, and Quadrant of Altitude, all curiously divided and lackered [these last two words deleted in MS], the like cannot be made again for Fifty Pounds: These Globes will be Raffled for on the 28th of this Instant March [deleted in MS and "Aprill" substituted] by Twenty Persons, each putting in Twenty Shillings, to be Thrown for with a Pair of Mathematical Dice, the most in Three Throws to have the Globes, or any other way, as the Major Part of the Company shall think fit. The said Globes are to be seen at J. Moxon's, at the Atlas in Warwick-lane, where they are to be Raffled for...."[11]

Moxon's globes were highly regarded and his shop was a well-known haunt. An anecdote of Aubrey's shows that young Edmond Halley as a schoolboy knew Moxon and visited his shop. Halley "went to Paule's school to Dr. Gale: while he was there

"printed and sold by Joseph Moxon," 1673. British Library, Bagford Collection, Harl. 5947, nos. 66 and 70. Earlier advertisements of 1653 link Moxon's name with that of John Sugar, globemaker. See E. G. R. Taylor, *Mathematical Practitioners*, p. 240.

10 Bagford Collection, Harl. 5947, no. 72.

11 Ibid., no. 64.

he was very perfect in the caelestiall Globes in so much that I heard Mr. Moxton [*sic*] (the Globe maker) say that if a star was misplaced in the Globe, he would presently find it."[12] Pepys also was a friend and satisfied customer. On 8 September 1663: he went "among other places, to Moxon's and there bought a payre of Globes, cost me 3 l. 10 sh.—with which I am well pleased, I buying them principally for my wife, who hath a mind to understand them—and I shall take pleasure to teach her. But here I saw his great Window in his dining-room, where there is the two Terrestriell Hemispheres, so painted as I never saw in my life, and nobly done and to good purpose—done by his own hand." On the evening of 21 October 1663 Pepys gave his wife her first lesson in arithmetic, "in order to her studying of the globes, and she takes it very well—and I hope with great pleasure I shall bring her to understand many fine things." On 2 December 1663, "to my office till 9 a-clock; and so home to my wife to keep her company, Arithmetique, then to supper and to bed. . . ." On 5 December, "and so home to supper and to bed—after some talk and Arithmetique with my poor wife. . . ." Next day, "she and I all the afternoon at Arithmetique; and she is come to do Addicion, Substraction and Multiplicacion very well—and so I purpose not to trouble her yet with Division, but to begin with the globes to her now."[13] As Pepys was learning his own multiplication tables only two years earlier, he too had made good progress. "A lecture to my wife in her Globes, to prayers and to bed" was still the order of the day on 14–15 January 1664. Pepys also ordered a pair of globes for his office. On 14 March 1664, "to Mr. Moxons and there saw our office globes in doing, which will be very handsome—but cost money." On 8 April he brought the office globes home, reporting them "done to my great content." Pepys paid Moxon on 29 April for the work he had done "for the office upon the King's globes." On that day a bill for over £20 was registered in the Navy Treasury for "new covering" a pair of globes at the Navy office.[14]

Moxon seems to have been the first globemaker in England to make pocket globes in cases bearing the celestial hemispheres, and several pairs of these survive (fig. 1). The larger Moxon globes are very rare. An 8-inch terrestrial globe preserved at Skokloster

---

[12] O. L. Dick, *Aubrey's Brief Lives*, pp. 120–121.

[13] Latham and Matthews, eds., *Diary*, IV (1971), 302, 343, 404, 406.

[14] Ibid., V (1971), 16, 83, 136, and editors' note, citing P.R.O., Adm.20/5, p. 281.

Fig. 1. Joseph Moxon's 3-inch terestrial pocket globe with its case bearing representations of the celestial hemispheres, c. 1670. British Library, Map Library, Maps C.4.a.4.(7).

in Sweden is one of the few, or perhaps the only one, known. Several examples are to be found of another globe that Moxon helped to produce, the so-called "English globe." This was made to the specification of Roger Palmer, earl of Castlemaine, whose invention was prompted by the sight of Moxon's 3-inch pocket terrestrial globe in its case. "Waiting upon my Lord in the beginning of Anno 1672, at his then arrival into England, I brought his Lordship (knowing that any thing new and ingenious would be acceptable to him) one of my 3 inch Terrestrial Globes, with the Stars described in the inside of its Case, which when his Lordship had considered, and bin inform'd by me, that its only Use was to keep in memory the situation of Countries, and Order of the Constellations and particular Stars, He intimated, that certainly much more might be done by it, and so returning beyond Sea fell upon this excellent Work. When he came home again (which happened above a year ago) and was pleased to shew me what he had done, I was as much ravisht and surprised at the admirable Contrivance of his Globe, and the many unexpected Operations performed by it, as if I had bin a new Beginner in the Study of the Sphere. . . . And this I can also say, that as to the Geographical part, it is (considering its bigness) not only the most useful, but also the best order'd and the best divided Globe extant; and yet it would have bin not a little better, had not his Lordships late Troubles hinder'd him from finishing it."[15] (His Lordship was then awaiting trial for high treason.) Moxon therefore constructed the globe himself, and published it in 1679, with an accompanying handbook entitled *The English Globe. Being a Stabil and Immobil one, performing what the Ordinary Globes do, and much more. Invented and described by the Earl of Castelmaine* (1679). The globe was fixed on a pedestal which carried at its base a dial marked with a stereographic projection of the stars. It was thus, in effect, a globe attached to an astrolabe, and was claimed to obviate the inconveniences met with in the use of Stoffler's astrolabe. "Nor, could my admiration be less [writes Moxon], when I saw how (without the usual assistance of a Meridian Line, Mariners Compass, or any such helps) it composed itself to the true site and position of the World."[16] Moxon's advertisement of it carried an illustration and a long list of the globe's advantages. The price with the projection was 50 shillings,

[15] Joseph Moxon, *The English Globe* . . . (1679), sig. A2r-v.
[16] Ibid., sig. A2v.

without it 40 shillings; and the book for its use was 5 shillings.[17] A testimonial to the excellence of the invention published in another of Moxon's advertisements was signed by six scholars of Oxford University and six of Cambridge.[18]

This invention was an ingenious device to represent the relative movement of the earth and the celestial sphere. It was famous in its day, for Father Vincenzo Coronelli, the great Venetian globemaker and cosmographer, gave a full description of it in his lectures at Venice, published in his *Epitome Cosmografica*, 1693. Nevertheless, the English Globe did not start any new fashion; the usual pair of terrestrial and celestial globes continued in production and use. Two examples of the English Globe survive in England, one in Winchester Cathedral Library, the other in the University Library, Cambridge.

In January 1662 Moxon petitioned the king "for the place of his Hydrographer, to make globes, maps, sea-plats &c. for accommodation of seafarers, having found the most perfect way yet known of making them; has also made in the English tongue the most exact and perfect Waggoner extant in any language, very necessary for seamen but fears it will not repay the cost, unless purchased by His Majesty." This petition was endorsed with the certificate of I. Newton, D.D., and thirteen others, chiefly professors of mathematics. Moxon was duly appointed, a warrant being issued on 10 January to Lord Chamberlain Manchester "to swear in Joseph Moxon as Hydrographer to the King, for making globes, maps, and sea-plats."[19] The "Waggoner" can be identified as *A Book of Sea-Plats, Containing the Scituation of all the Ports, Havens . . . in all Europe. Newly Corrected by Joseph Moxon . . . 1657* (fig. 2).[20] It was the first sea-atlas published in England during that period, and marked England's entry into a field hitherto monopolized by the Dutch. The charts themselves appear to be copies made from Dutch originals, with titles and legends translated into English. The prototype seems to have been an early sea-atlas of Pieter Goos, whose first pilot-book, *Lichtende Columne afte Zee-Spiegel*, was sold in 1650. The fact that the imprint of the chart of Russia, Lapland, and Finmarck refers to Moxon's "schop" suggests that the engraver

17 Bagford Collection, Harl. 5947, nos. 61, 63.
18 Ibid., 5946, no. 203.
19 *Calendar of State Papers, Domestic Series 1661–1662* (London, 1861), p. 241.
20 British Library, K. Mar.1.(41). A second edition published in 1665 is in the Bodleian Library. (Maps 2012,b.4).

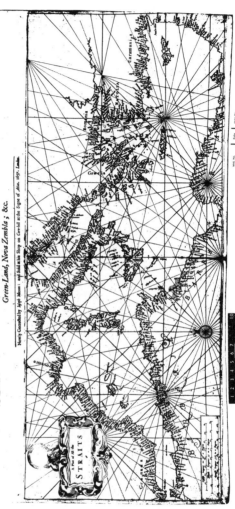

A BOOK OF SEA-PLATS.

Containing the Scituation of all the *Ports, Havens, Greeks, Capes, Rocks, Sands*, and *Shoalds*, in all

EVROPE.

Whereby is shewn the *Distance* of Places one from another, and upon what Point of the *Compass* they bear; as also the *Depths* or *Soundings* upon each respective Coast.

Beginning at the *Straits* ; and from thence passing all along the whole Coasts of *Portugal, Spain, France, Flanders, England, Scotland, Ireland, Holland, Denmark,* all the *East Sea, Norway, Finmark, Lapland, Russia, Green-Land, Nova Zembla ;* &c.

Newly Corrected by *Ioseph Moxon* : and Sold at his Shop on *Cornhil* at the Signe of *Atlas.* 1657. *London.*

Fig. 2. Moxon's "Waggoner," *A Book of Sea-Plats*, 1657; first page, displaying the title, with a chart of the Mediterranean engraved below. The word "Europe." is in MS. British Library, Map Library, K.Mar.I.41.

may have been Dutch. Decorative vignettes on the charts are etched in a style characteristic of Hollar, and different from that found in Dutch atlases. Moxon probably acquired the original Dutch work on his visit to Amsterdam in 1652.

In addition to globes and charts, Moxon was making and selling a wide range of mathematical instruments with booklets on their use. These instruments were of the cheaper type, the lines and scales printed on paper and pasted on boards, as he mentioned in his advertisements.[21] He was not in the class of mathematical practitioners who worked in metal, for his instrument making was essentially an adjunct to his work as a printer. As a master printer who was also a globemaker and chartmaker, he combined an unusual range of activities. In the first volume of his now celebrated *Mechanick Exercises, or the Doctrine of Handy-Works* (1677–1680), Moxon describes the trades of artisans: metalworkers, cabinetmakers, house carpenters, and woodturners, and in the second volume, published in 1683, printing.

This comprised the first full description of the printing trade. Published in parts, the *Exercises* now ranks as one of the most celebrated "periodicals" of the day. Moxon opened the preface as follows: "I see no more reason why the sordidness of some Workmen should be the cause of contempt upon Manual Operations than that the excellent invention of a Mill should be despised, because a blind Horse draws in it. And though the Mechanicks be by some accounted ignoble and scandalous. Yet it is very well known, that many Gentlemen in this Nation of good Rank and high Quality are conversant in Handy-works. And other Nations exceed us in numbers of such. How pleasant and healthy this their Divertion is, their Minds and Bodies find; and how harmless and honest all sober men may judge. That Geometry, Astronomy, Perspective, Musick, Navigation, Architecture &c. are excellent Sciences, all that know but their very names will confess: Yet to what purpose would Geometry serve, were it not to teach Handicrafts? Or how could Astronomy be known to any perfection, but by Instruments made by Hand?"

21 For example, in the advertisement in J. Newton, *A Help to Calculation* (London, 1657), sig. Q 1ᵛ. Listing Blagrave's, *Mathematical Jewel,* Oughtred's, *Circles of Proportion*, etc., he adds: "These, with divers other Mathematical Instruments, are printed on Papers, and exactly pasted on Boards, &c with Indexes, and other appertenances proper to each Instrument, very commodious for any mans use." Moxon's "sea-plats" also were described as "printed on paper or parchment, and pasted on Boards."

This emphasis on practical application was a constant theme of the writers and instrumentmakers of the day. What was more practical or essential, moreover, than the art of the Smith with which the *Exercises* opened, for (as Moxon remarks) on this depended all other trades. Hooke writes in his diary for Monday, 31 December 1677: "Calld on Moxon, he read me his first monthly exercise of smithery and preface in order to license," and on Monday, 7 January, "Bought of Moxon his 1st monthly exercise, 6d."[22]

On 30 November 1678 Moxon received the greatest honor of his professional career in being elected a member of the Royal Society, in company with his young friend Edmond Halley.

Moxon's rivals and successors were Robert Morden, William Berry and Philip Lea—globemakers who worked separately and in partnership. Described as being "although industrious, always in pecuniary difficulties,"[23] Robert Morden was in business at the Atlas in Cornhill from 1668 until 1703. The *Term Catalogues* (10 July 1671) show that with William Berry, he too was selling 3-inch terrestrial globes with cases bearing the celestial hemispheres, "of great use for all Gentlemen, or others. . . ." The notice also mentions "a new Size of Globes ten inches Diameter; also all sorts of Globes, Spheres, Maps large and small, Sea-Platts, and other Mathematical Instruments."[24] An advertisement in the Bagford Collection invites subscriptions "For a New large Size of Globes, By Robert Morden at the Atlas in Corn-Hill, and John Rowley, Mathematical-Instrument-Maker, under S. Dunstan's-Church in Fleetstreet." The publishers promised "Stars, Cities, &c. gilded . . . Whosoever shall be pleased to subscribe . . . shall have his Coat of Arms, Name, Title, Place of Residence, &c., inserted, on some convenient Place on the Globes, if desired." Each subscriber was to pay the sum of £10 to Mr. Chambers, goldsmith, near Temple-Bar in Fleetstreet, and £10 more upon delivery. The undertakers promised not to sell any pairs but to subscribers, under £25 the pair.[25]

Several examples survive of the 14-inch globes that Morden, Berry, and Lea made in partnership in about 1683 and were selling at their three shops. The address to the reader (on a slip

---

[22] H. W. Robinson, and W. Adams, eds., *The Diary of Robert Hooke . . . 1672–1680* (London, 1935), pp. 337–338, 339.
[23] *Dictionary of National Biography.*
[24] Edward Arber, *The Term Catalogues I* (London, 1903), 80–81.
[25] Bagford Collection, Harl. 5946, no. 204.

pasted over the cartouche on one now in the Whipple Museum, Cambridge) claims "there is not any part of ye Earth wherein we have not made a considerable alteracion." The tracks of Drake and Cavendish are marked, as also on Palmer's "English Globe." The terrestrial appears in the portrait (now at the National Maritime Museum) of Captain John Benbow, master of the fleet under Admiral Russell at the Battles of Beachy Head, Barfleur, and La Hogue, 1692, with (to the right) Sir Ralph Delavall, vice-admiral of the Red Squadron. Pepys probably did not obtain a new pair of globes from Morden, Berry, and Lea, for he was making enquiries in December 1687 about Blaeu globes, seeking "the best Globes to be bought in Holland." His informant, Mr. Faseley, replied: "the finest Globes being 26 inches diameter with Fannered frames they cannot be Red[y] Befour a months warning being Given. Price 250 Guilders, in English money £22.14s.6d. The ordinary Globes of 26 inches diameter with waincot frames these cane be Redy in a weeks time after warning given. Priced 150 Guilders in English money £13.12.od."[26] The globes were available in Amsterdam from Joannes van Keulen, who had taken over the stock of Joan Blaeu after his death. It appears that Pepys did not complete the transaction, perhaps because of his fall from favor in 1688. At £22.14.6 for the finest pair, the price of the Blaeu globes was comparable to that of Moxon's at £20 for 26-inch globes, and Morden's at £20 to £25 for the 30-inch globes. The ordinary pair in wainscot frames at £13.12.0 was unusually cheap. Probably these comprised only the spheres mounted in protective timber for transit.[27]

Morden obtained the patronage of Thomas Goddard of London, merchant, for an important venture in geographical publishing. In his *Geography Rectified: or, a Description of the*

[26] See Arthur Bryant, *Samuel Pepys The Saviour of the Navy* (Londin, 1949), p. 224. The original document is in the Bodleian Library, (Rawl. MSS, A.171.f4), and is quoted by permission of the curators of the Bodleian Library. Tony Campbell in "A Descriptive Census of Willem Jansz. Blaeu's sixty-eight centimetre Globes," has investigated the meaning of the terms "fannered" and "waincot." Informed by the Victoria and Albert Museum that "fannered" means veneered (from fanér in Swedish), he points out, however, that no veneered work has been observed in the stands of surviving Blaeu globes. "Wainscot" is defined as "a superior quality of foreign oak" (OED) *Imago Mundi*, 28 (1976), 32–33. For Blaeu's globes see C. Koeman, "An Inventory of Johannes van Keulen's Globe-Factory in Amsterdam, dated 1689" (Paper presented to the IIIrd International Symposium on the History of Cartography, Brussels, 17–20 September 1969); also C. Koeman, *A Catalogue of Joan Blaeu* (Amsterdam, 1967), p. 6.

[27] An interpretation suggested by Tony Campbell.

World (1680), he set out to present "such a satisfactory view of the Earthly Globe . . . as may make good our Title," and to correct the current errors in geographical observations and maps. Referring to his "many years Experience not only in making and projecting of Globes, Maps &c., but also in examining and comparing of the Discoveries, Observations, Drafts, Journals and Writings, as well of the Ancient as Modern Geographers, Travellers, Mariners, &c.," he admitted that his work "wants the helps and advantages of a more Learned Pen, and in Truth it ought to have been freed from those frequent avocations and disturbances that attend a Publique Shop and Trade."[28] Morden's *Geography* proved to be a very popular and successful handbook. Well received at Oxford and Cambridge (as noted in the second edition), it ran to four editions by 1700, with another, described as the "Fourth," by "R. R.," also published in 1700. In terms of critical method, the work was not in the same class as the great treatise by Father J. B. Riccioli, *Geographia Reformata*, which undertook a systematic analysis of observations, but whose voluminous scholarship was accessible only to scholars well versed in Latin.[29]

As mapsellers, Morden and Berry were two of the most prolific of their day, with a wide range of publications to their name. Among the visitors to Morden's premises in 1676–1677, was Robert Hooke, accompanied by Sir John Hoskins, the future president of the Royal Society. They bought maps by Pierre du Val, and on a later occasion, examined there a new map of Alsace.[30] William Berry at the Globe in Charing Cross was well known to Pepys, who minutes: "Speak to Mr. Bury [*sic*] the globe-maker to be informed what sea-maps have been made in France."[31] Berry was the man to answer such a question, for (as Pepys notes), "Take notice of Mr. Bury's maps done in England in imitation of Sanson's French."[32] Philip Lea, at "Ye Atlas & Hercules in Cheapside, London," was Pepys's professional consultant on maps and globes. "Globes, when and where first in-

---

[28] Robert Morden, *Geography Rectified: or, a Description of the World* (London, 1680), "To the Reader," sig. A3ᵛ–A4ᵛ.

[29] Morden's work in terms of so-called "critical Geography" is assessed by Edward Cave in his *Geography Reform'd* (1739), pp. 252–244.

[30] Robinson and Adams, *Hooke's Diary*, 15 May 1676, 1 August 1677, pp. 232, 304. "Mordants" (p. 232) is presumably a mistake or misreading for "Mordens."

[31] Tanner, ed., *Pepys's Naval Minutes*, p. 122.

[32] Ibid., p. 120. Another entry reads "*Memorandum.* That Sanson's maps and tables gave the first hint to our Mr. Bury [*sic*] to undertake his in the same manner" (p. 177).

vented, and when first here? Consult Mr. Lee [*sic*], etc., thereon," he writes in his *Naval Minutes*.[33] Lea stained and colored Pepys's maps.[34]

A globemaker in business at the turn of the century was the instrument maker Thomas Tuttell, whose name appears with James Moxon (Joseph Moxon's son) on the title page of the fourth edition of *Mathematicks made Easie* (1705). With Col. William Parsons, who is named as his patron, Tuttell was advertising a new pair of 36-inch globes, accompanied by a treatise on their use, costing £25 a pair, one guinea for each globe to be paid on entry and the remainder on delivery. This advertisement and another illustrating all the instruments Tuttell had on sale, including the globes, are preserved in the Bagford Collection (fig. 3).[35] It is interesting to note that a celestial globe by Tuttell, 14 inches in diameter, similar to one in the advertisement, recently appeared in the London salerooms. Unlike Moxon, Tuttell worked in metals and other nonpaper materials, advertising instruments in "Silver, Brass, Ivory and Wood."[36] He was appointed instrument maker to the king in 1700, and numbered Pepys among his customers.[37]

Tuttell and Parsons may have been inspired to produce (or offer) 36-inch globes in emulation of Father Coronelli, whom they mention in the advertisement of their treatise. The indefatigable Coronelli had come to London from Venice in 1696, visited the Royal Society, and on 11 May presented to King William III a pair of 1½-foot globes dedicated to the king for which he was rewarded with 200 gold guineas. (A pair of these globes is preserved in the National Maritime Museum.) On the sixteenth of May, Coronelli was in the company of Hooke and Halley observing a total eclipse of the moon and the satellites of Jupiter. In the same month he visited the universities of Oxford and Cambridge and the Royal Society in London, which honored him with membership.[38] Coronelli's most celebrated printed

[33] Ibid., p. 419.

[34] J. R. Tanner, ed., *Private Correspondence and Miscellaneous Papers of Samuel Pepys 1679–1703* (London, 1926), I, 166–167; II, 328–329.

[35] Bagford Collection, Harl. 5946, no. 205; Harl. 5947, no. 88 (illustration of instruments). An advertisement of Tuttell's instruments also appears in *Mathematicks made Easie* (1705), B.L. 1489.g.29.

[36] Bagford Collection, Harl. 5947, no. 77.

[37] Tanner, ed., *Private Correspondence*, I, 167. (Tuttell's name appears in a list of "Debts," along with Lea, Thornton and other dealers.)

[38] Vincenzo Coronelli, *Viaggi del P. Coronelli* . . . [*Viaggi d'Italia in Inghilterra*], II (1697), 154, 204. See also *Il P. Vincenzo Coronelli* . . . *nel III Centenario della nascita* (Rome, 1951), pp. 251–252.

Fig. 3. Thomas Tuttell's advertisement of globes and other instruments for sale. British Library, Bagford Collection, Harl. 5947, no. 88.

globes, which he claimed to be the best ever made, were his 3½-foot globes. First published in 1688, they were reduced versions of the great 15-foot MS globes made for and presented to King Louis XIV in 1683, by which Coronelli had established his reputation as a globemaker. He organized, in Paris and elsewhere on the continent, subscription societies for the purchase of his atlases, maps, and globes. These were run under the auspices of the Cosmographical Society of the Argonauts (Academia Cosmografica degli Argonauti), which he founded in Venice in 1684, and which ranks as the first geographical society on record. The Bagford Collection includes a subscription form for the purchase of Coronelli's maps and globes in London.[39] John Cailloué, French bookseller at the corner of Beauford-Buildings in the Strand, was prepared "if he meets with any Encouragement," to procure 200 atlases and 100 globes from Venice, "for such as will be pleased to Subscribe under the following Conditions. . . ." The celestial and terrestrial globes 3½ feet in diameter, ready made up, were priced at £30. The "two foot Globes" (1½ foot?) were priced at £10. Neither the *Atlante Veneto* nor the globes were to be transported out of His Majesty's dominions during the year 1697, under forfeiture of the globes or atlas so transported, in order that those who had undertaken the same for other countries might not be prejudiced. That the globes could only be mounted in London was probably agreed upon during Coronelli's visit in 1696. He was finding it difficult to substantiate his claim to have produced the finest globes yet made, when so few examples were available for inspection in the northern countries of Europe, especially in England. Hence his production of a "book of globes," the *Libro dei Globi* published in 1697, with later editions up to about 1710.[40]

John Seller (fl. 1669–1691), "compass-maker" and mapseller, was another leading mathematical practitioner in London.[41] Setting up in the 1660s at the Sign of the Mariner's Compass, Wapping, for forty years he carried on a nautical business unrivaled in London. The title of king's hydrographer, which he shared with Moxon, was conferred on him in 1671,[42] and was

[39] Bagford Collection, Harl. 3947, no. 86.

[40] See Helen Wallis, ed., *Vincenzo Coronelli Libro dei Globi* . . . (Amsterdam, 1969), p. xii.

[41] See especially the fourth essay in this volume, by Coolie Verner, pp. 127–157.

[42] *Calendar of State Papers, Domestic Series*, January to November 1671 (London, 1895), p. 144, entry for 24 March 1671.

held through three reigns. He is claimed to have been the first tradesman not only to list and advertise the instruments he made and sold, but also to be prepared to teach the use of them. The full list of his stock as advertised in *The Term Catalogues* for 7 February 1671/72 includes a wide range of mathematical instruments of all kinds, including "Globes with Sphears of the Heavens, etc. . . . Maps of the World of all sizes, and of any particular Country; with any other Mathematical Instruments whatsoever."[43] A later catalogue pasted at the back of his *Atlas Maritimus* (1675)[44] lists his stock in about 1684 or 1685.

Preeminent as a nautical mapmaker and publisher, Seller embarked on one of the most ambitious projects of the day with his undertaking of *The English Pilot*, announced in 1669.[45] This was hailed as the first English sea-atlas to appear since the publication of the "Waggoner" in 1588 (although strictly speaking, Moxon's *Book of Sea-Plats* (1657) predated *The English Pilot*). Waghenaer's *Mariner's Mirror*, translated from the Dutch by Anthony Ashley, and popularly known as "The Waggoner," with the later Dutch editions, was still an accepted authority, as Pepys notes: "the subsequent editions of the Dutch Waggener, translated and printed by the Dutch themselves in English, with their own Dutch plates, have from time to time to this prevailed among us English . . . till Seller fell to work."[46] Pepys was acquiring for his collection copies of the "Waggener" as well as Seller's new atlases and charts.

Seller's new English sea-atlas had, however, a serious defect. Its maps were worked up from the old copper plates of Dutch charts. Pepys minutes: "Seller's maps are at the best but copies of the Dutch, with such improvements as he could make therein by private advice upon the observations of single men,"[47] and again—"They say that even Seller's new maps are many of them little less than transcripts of the Dutch maps, some of them even

[43] *The Term Catalogues*, I, 100.

[44] John Seller, *Atlas Maritimus* (1675), British Library (Maps C.8.d.5).

[45] See Coolie Verner's essay "Engraved title-plates for the folio atlases of John Seller," in Helen Wallis and Sarah Tyacke, eds., *My Head is a Map: A Festschrift for R. V. Tooley* (London, 1973), pp. 21–52.

[46] Tanner, ed., *Pepys's Naval Minutes*, pp. 348–349. The plates of Waghenaer's *Mariner's Mirror*, published in England in 1588, had been removed to the Netherlands, and were reprinted only once, in 1605, with Dutch text, by Jodocus Hondius at Amsterdam.

[47] Tanner, ed., *Pepys's Naval Minutes*, p. 135. See also the essay in this volume by Jeanette Black.

with papers pasted over and names scratched."[48] Seller had bought worn Dutch plates for old copper and refurbished them. New surveys and new maps in fact demanded resources and capital which were beyond the means of successful mapdealers. Seller obtained sales even though his wares were secondhand. A contemporary wit wrote:

> This of your Book, and you, I may foretell,
> What SELLER made, be sure will SELL.[49]

The privilege granted by Charles II to Seller in 1671 gave him protection for his sea-atlases, "forbidding any person to print any work, under any title, reprinting or counterfeiting, for thirty years, the works of John Seller, the *English Pilot* and the *Sea Atlas*, describing the coasts, capes, headlands, &c., of the kingdom: also forbidding the import from beyond seas of any such books or maps, under names of the *Dutch Waggoner or Lightning Column*, or any other name."[50] Nevertheless Seller's ambitions exceeded his resources. Faced with the threat of bankruptcy in 1677, he went into partnership with John Thornton, William Fisher, James Atkinson, and John Colson for the further publication of *The English Pilot*. When the partnership was disbanded in 1679, the title to most of *The English Pilot*, and the *Atlas Maritimus* passed to William Fisher.

There was thus in London in the 1670s and 1680s a vigorous intermingling of trades and crafts. Some instrument makers were also printers, some were engravers (it was from engravers amongst other trades that the mechanical instrument making craft had developed, as Professor Taylor points out).[51] Globe-makers were sometimes instrument-makers, sometimes printers, and sometimes both; some printers were also booksellers. In 1700 London had 42 printers, 188 booksellers (or publishers), and 25 men who practiced both trades.[52] Of mapsellers proper, 35 were active between about 1650 and 1710, and, at any one

---

[48] Ibid., p. 345. An earlier note by Pepys is also of interest: "Get Gascoin the plat-maker to compare the original and latter Waggeners with Anthony Ashley's and Seller's new maps" ibid., p. 42. Joel Gascoigne was one of the so-called Thames School of chartmakers. See T. R. Smith's essay in this volume.

[49] "To his worthy Friend, the Author, on his Elaborate and Useful Work, a Pindarick Ode." *The English Pilot; The First Book* (London, 1671), sig.b.2ʳ.

[50] *C.S.P. Domestic Series*, January to November 1671 (1895), p. 140, entry for 22 March 1671.

[51] E. G. R. Taylor, *Mathematical Practitioners*, p. 162.

[52] C. Plant, *The English Booktrade* (London, 1939), p. 64.

time, between 13 and 19 were in business.[53] Seller, Morden, Berry, and Lea commanded a large share of the map market. Lea, William Morgan, James Moxon, and Seller had stalls in Westminster Hall as additional outlets for map selling (fig. 4).[54] On the debit side, many of the products of the printing trade were of poor quality, and plagiarism was rife. Berry's maps were copies from the French—his best-known atlas was called "the English Sanson." Seller's charts were copies from the Dutch, and the London publishers in turn found their works were being copied and warned subscribers to beware of counterfeits.[55] Nevertheless the mapsellers and printsellers were meeting an insatiable popular demand. The center of the print trade was the City, where Pepys went on frequent visits, "went to the printed pictureseller's in the way thence to the Exchange; and there did see great plenty of fine prints. . . ." (18 April 1666).[56] The stationers in the graphic arts were free from company rules, without guilds, and could be a law unto themselves.[57] The use of subscriptions and advertisements to sell their wares provides useful supplementary evidence of publishing projects, some never completed, some never begun. The pages of the *London Gazette* which carried their advertisements are one major documentary source.[58] Another source is the collection Pepys's friend John Bagford, shoemaker and collector, put together for his history of printing. This collection was later acquired by the Harleys and came to the British Museum in the Harleian Library.[59]

The prints and maps produced in such profusion provided men like Pepys and Bagford with exceptional opportunities for collecting. For one shilling Pepys could buy a mezzotint of Charles I, or a map of France, and for the same sum he could have dinner at the King's Head. The new size of globes by Morden and Berry

[53] Sarah Tyacke, "Map-sellers and the London map trade c. 1650–1710," in Wallis and Tyacke, *My Head is a Map*, pp. 62–80, especially pp. 67, 80.

[54] Ibid., p. 75, plate 12.

[55] Such warnings of counterfeits were given by John Seller, as stated in King Charles II's privilege: *The English Pilot; the First Book*, sig.al[r]. (Maps 22.d.1); and also by John Ogilby, with reference to his map of London and Westminster, *London Gazette*, no. 1888, 20–24 April 1676.

[56] Latham and Matthews, eds., *Diary*, VII (1972), 102.

[57] L. Rostenberg, *English Publishers in the Graphic Arts 1599–1700* (1963), pp. 1–4, 75, 93.

[58] These are to be reproduced in a forthcoming publication by Sarah Tyacke, *London Map-sellers', 1660–1720* (London, 1977 [–78]).

[59] The following volumes of the Bagford Collection are of special interest for their details of cartographic projects: Harl. 5935, 5946, 5947, 5956, 5957, and a volume comprising 5906B, 5908, 5910 (Pt. 1–4), 5943, 5953 (Pt. 1), 5997.

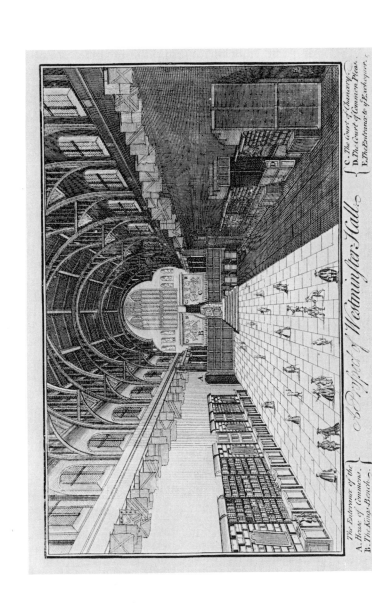

The Entrance of the { A. House of Commons. B. The Kings-Bench. } ~A Prospect of Westminster Hall.~ { C. The Court of Chancery. D. The Court of Common Pleas. E. The Entrance to the Exchequer. }

Fig. 4. A Prospect of Westminster Hall, engraved c. 1690, showing booksellers' and mapsellers' stalls. Philip Lea's stall, which stood next to the Court of Common Pleas (c. 1689–1695), displays a map of England and Wales. British Library, Map Library, K.Top.XXIV.24b.

at £4 equaled the annual wage of Pepys's new Cooke-maid "at
4 l per annum, the first time I ever did give so much—but we
hope it will be nothing lost by keeping a good cook" (26 March
1663).[60]

For all the variety and vigor of English publishing and printing
in the graphic arts, large scale projects for publishing atlases were
a formidable proposition. Various ambitious schemes were proposed but failed to reach completion as the entrepreneurs of
these projects overstretched themselves. Thus John Ogilby, who
started as a dancing master and was to be appointed in 1671 His
Majesty's cosmographer, planned his *English Atlas*, as he described it, "girding himself couragiously for no less than the
Conquest of the whole World, making the Terrestrial Globe his
Quarry, by a New and Accurate Description of its four Quar-
ters."[61] The atlas project was to be financed by subscriptions
that Ogilby had already used to good effect. His map of London,
1682, published with William Morgan, shows him presenting to
the king his list of subscribers to the earlier version of 1677. He
succeeded in publishing the volumes for Africa (1670), and
America (1671), the title page of which proclaimed him as "His
Majesty's Cosmographer, Geographick Printer, and Master of
the Revels in the Kingdom of Ireland"—followed by Asia in
1673. Of the *Britannia* only one volume, the famous road atlas,
had appeared by 1675. The cost of survey had run to some
£1,400. The king headed the list of subscribers willing to meet
the expenses, proposing "our Royal Example, by subscribing
Five Hundred Pounds," and adding £500 more on behalf of His
Royal consort; and solicited "cherefull subscriptions and contributions of all and every the Lords ArchBps. & Bpps. Dukes
Marquesses & Earls Viscounts Lords Lieutenants & Deputy
Lieu^ts: both ye Universities, the heads of Collidges & Halls therein," as well as ecclesiastical, judicial, and military authorities.[62]

[60] Latham and Matthews, eds., *Diary*, IV (1971), 86. L. Rostenberg, *English
Publishers*, makes this comparison (p. 93).

[61] "A Proposal concerning an English Atlas," dated 10 May 1669, Bodleian
Library, Wood 658, fol. 792; quoted in Sarah L. C. Clapp, "The Subscription
Enterprises of John Ogilby and Richard Blome," *Modern Philology* 30 (1933),
368. See also J. B. Harley, ed., *John Ogilby: Britannia, London 1675* (Amster-
dam, 1970), p. ix, on Ogilby's royal warrant and appointment as His Majesty's
Cosmographer (Bodleian, MS Aubrey 4.f.244).

[62] Bagford Collection 5946 no. 166 (MS Memorandum), also Royal Procla-
mation of 11 July 1672, ibid., no. 178. The cost of the survey is given in the
prospectus for the *Britannia*, c. 1672, a document in Yale University Library,
cited by Harley (1970), pp. ix–xi.

As a further financial aid, subscriptions having failed to produce enough money, Ogilby organized a standing lottery, a device he had already used for publishing *An Embassy from the East-India Company ... to the Grand Tartar Cham, Emperor of China* (1669). A reference to his "excellent invention and prudential wit" in the field of lotteries and subscriptions implies that he was considered the originator of both these financial devices in publishing.[63] The lotteries of the Royal Fishery Company were, of course, a well-established feature of the day, and Ogilby's lotteries were licensed by the Master of the Revels, His Royal Highness the duke of York and assistants of the corporation of the Royal Fishing. It may perhaps have been Ogilby's experience as Master of the Revels in Ireland (1662–1665) which gave him the idea of applying lotteries to publishing. The lottery of his own previously published works, first arranged for 1665, but stopped because of the plague (Pepys was successful in one of the draws), was taken up again in 1668, to repair the losses suffered in the Great Fire. Notices are to be found in the *London Gazette*, and detailed proposals in the Bagford Collection.[64] Advertised on 14–18 May for 25 May 1668, it was postponed until 2 June, "Adventurers coming in so fast that they cannot in so short time be methodically registered." In 1676 when Ogilby died, his partner William Morgan, who was requested in Ogilby's will to finish the *Atlas*, organized the sale of his books by auction.[65] On William Morgan's death, Robert Morden in his turn purchased the remaining stock and organized a lottery in March 1690/91,[66] an event advertised in fine satirical style in an engraving by Sutton Nicholls (fig. 5).[67]

Another project which foundered was Moses Pitt's similarly named "English Atlas." (Pitt probably chose the name after Ogilby's death in 1676.) This was the first large general atlas to be launched in England and was conceived (as its proposals state) on the model of the great eleven volume Blaeu and Jansson atlases. In emulation of these, the *English Atlas* was to comprise eleven volumes, but it fell far short of the promised number.

[63] Earlier short-lived ventures of 1615, 1617 and the 1650s are noted by S. L. C. Clapp, "The Subscription Enterprises," 365–366; see also p. 370.

[64] *London Gazette*, no. 260, 11–14 May 1668, repeated 25–28 May 1668; Bagford Collection, Harl. 5946, nos. 179, 180; Bagford Collection, Harl. 5947, no. 2 (plate).

[65] J. B. Harley, ed., *John Ogilby: Britannia, London, 1675*, p. xviiii.

[66] Proposals in the Bagford Collection, Harl. 5946, no. 186.

[67] Ibid., 5947, no. 2.

Fig. 5. "The Compleat Auctioner," an engraving by Sutton Nicholls to advertise the sale of John Ogilby's Library [c. 1691?]. British Library, Bagford Collection, Harl. 5947, no. 2.

Only four volumes had appeared when the project had to be abandoned as a failure. The task was beyond Pitt, who was neither a scholar nor a cartographer.[68] The gilded preliminaries of the atlas, its grandiloquent address to the king, its impressive list of subscribers (including the names of Pepys, Hooke, Evelyn, Jonas Moore, John Wallis, and Sir Christopher Wren, and a host of others) show the generous support offered by men of letters and men of affairs. Robert Hooke declared "His design for Atlas good." A committee of the Royal Society was set up to supervise the publication, and, finally, Hooke took over the main task of supervision for a fee of £200 which he never received from "that Rascal Pitt." The maps themselves were printed from Jansson's copperplates that Pitt obtained by contract with Jansson's heirs Swart and Van Waesberge. Only one map was designed, drawn, and engraved in England, the map of the North Polar regions, which bears the imprint "Oxford, 1680," and clearly was derived from Seller's map. "Mr. Nott," writes Pepys (referring to the famous London bookbinder William Nott), "tells me that Mr. Pitt's new atlas has the very Dutch words and Dutch arms not taken out of the maps, and [the] D[uke] of Y[ork]'s taking notice to Pitt (when he presented it to him) of the maps being of too small and close a letter, who answered, Yes, but that they had a very fair margin."[69] The arrest of the luckless Pitt in 1658 and his later imprisonment for a dubious venture in the London property market brought the project to an end.

Seller also became the entrepreneur of a so-called English atlas, this one with the latinized title *Atlas Anglicanus*. It was intended as a folio county atlas, to be produced in collaboration with the surveyor John Oliver and the engraver Richard Palmer. On 15 September 1679 the king referred to the Lords of the Treasury the petition of John Seller, John Oliver, and Richard Palmer:

shewing that they have undertaken a great and elaborate work of an actual survey of England and Wales, and to comprise the same on complete maps of every county in a large book in folio, entitled *Atlas Anglicanus*, and praying liberty of importing customs free 10,500 reams of elephant paper for printing same, as it is a work of

[68] See E. G. R. Taylor, " 'The English Atlas' of Moses Pitt," *Geographical Journal* 95 (1940), 292–299.

[69] Tanner, ed., *Pepys's Naval Minutes*, p. 105.

very great expense and never heretofore effectually performed by any.[70]

The king directed a payment of money rather than a customs free grant, as appears from a reference to Seller against a similar petition from Moses Pitt for his Atlas;[71] and a royal bounty of 200 l. was granted to Seller on 10 March 1679/80.[72] Pepys must have been referring to this in his minute, "Very good matter to be found in the King's grant to Seller's new undertaking, 1679."[73] A preliminary version of Seller's atlas with MS title page is preserved in King George III's Topographical Collection (now in the British Library), and presumably came from the Old Royal Library. It was probably presented by Seller to King Charles II as a model of the projected atlas. A second example, with printed title page, and maps by a variety of cartographers, turned up in the London salerooms in 1969.[74] The preparation of this "atlas factice" (i.e., an atlas made up by a mapseller from map sheets in his stock) suggests that Seller was now trying to win for such atlases a place in the market, which was already being profitably exploited by Overton and by Lea. The project for the *Atlas Anglicanus* never came to fruition, and by 1693 the partnership with Oliver and Palmer had been dissolved.

The lack of original survey, the expense of supplying it, the cost in capital outlay of making new copperplates for a major atlas—these factors explain why the various publishers of the day, Overton, Lea, Seller, and others, produced this type of "atlas factice," using maps and plates from other publishers as well as any original maps of their own that came to hand. If a gentleman or official needed a set of maps, he would commission a mapseller

[70] *Calendar of State Papers Domestic, January 1st, 1679, to August 31st, 1680* (1915), p. 242; *Calendar of Treasury Books, 1679–80*, vi (1913), 276–277 (under 29 November 1679).

[71] "Report to the King from the Treasury Lords on the petition of Moses Pitt, bookseller, who prays for the importation of such paper, maps and books from beyond the seas, Customs free, as he shall have occasion of for carrying on his Atlas and other undertakings. Great inconvenience and trouble may arise to the Customs by such a grant. What your Majesty shall be pleased to bestow upon him will be best paid in money 'as you were pleased to direct in the like case of John Seller and partners'"; *Calendar of Treasury Books*, vi (1913), 718–719, entry for 27 October 1780.

[72] Ibid., p. 678.

[73] Tanner, ed., *Pepys's Naval Minutes*, p. 223.

[74] R. A. Skelton, *County Atlases of the British Isles 1579–1703* (London, 1970), pp. 187, 217–218.

to supply the maps and have them bound. This presumably explains the origin of an atlas, more properly described as a collection of maps, c. 1690–1700, lettered on its spine "English Plantations in America." Now to be found in Worcester College, Oxford, the atlas was presented by George Clarke (1661–1736), secretary at war in Ireland 1690–1692, secretary at war in England from 1692–1693 to 1704, and for several years (1702–1705), secretary to Prince George of Denmark, Queen Anne's husband. Another "made-up" atlas of a different kind is the Blathwayt Atlas, comprising a collection of manuscript and printed maps put together by William Blathwayt as secretary to the Lords of Trade and Plantations.

One other atlas worthy of mention, like the *English Atlas* of Moses Pitt, bore the imprint "At the Theater in Oxford." Entitled *A New Set of Maps* by Edward Wells, student of Christ-Church, with accompanying text *A Treatise of Antient and Present Geography* (1701), the work was designed for the use of young students at the universities. Its didactic purpose was rigorously interpreted and explains the stark appearance of the maps, for both maps and text were designed to be memorized. "It was judg'd proper to let nothing have a Place either in the one or the other, but what should deserve likewise a constant Place in the memory."[75] Thomas Hearne the antiquary while commending Wells's industry, declared his books inaccurate and containing "very little that is curious."[76] The interest of the publication lies in the fact that geography had gained an accepted place in the Oxford syllabus. That mathematics also was now a school and university subject, and not merely part of technical training, is indicated by another of Wells's many publications, his textbook *The Young Gentleman's Course of Mathematicks* (1712–1714), which like the *Geography* ran to a number of editions. Mathematics also had been freed from the restrictions imposed by use of Latin and Greek. As Aubrey pointed out in 1690, "the Mathematicks" had become the most popular of all studies: "it was Edmund Gunter who, with his Booke of the Quadrant, Sector and Crosse-Staffe did open men's understandings and made young men in love with that Studie. Before, the Mathematicall Sciences were lock't up in the Greeke and Latin

75 E. Wells, *A Treatise of Antient and Present Geography* (1701), sig. b^r.
76 T. Hearne, *Remarks and Collections of Thomas Hearne*, ed. H. E. Salter, ix (London, 1914), 330.

tongues and there lay untoucht, kept safe in some libraries. After Mr. Gunter published his Booke, these Sciences sprang up amain, more and more to that height it is at now (1690).''[77]

Thus in the later years of the seventeenth century, facilities for geographical and mathematical instruction were available both in the shops of London dealers and at the universities. Thomas Tuttell offered instruction at his two shops in "all parts of the Mathematics" (including the use of instruments). A typical advertisement appearing in the Bagford Collection proclaims "Geography made Easy, and the Use of all the ordinary Sorts of Charts and Maps, whether Geographical, Hydrographical, Plans, Groundplots, or Perspectives, Taught in a Week's time; ... And an easy Explication of the hard Words, which may discourage some People from this Necessary, Pleasant and Easy Science. It is Taught to either Sex, whether Learned in other Sciences or not, if they be above the Age of twelve years. The Master Teaches either in his own Chamber, or comes to the Scholars. . . . The Master may be heard of at *Mr.* Bell's *Bookseller at the* Bible *and* Cross-Keys *in* Cornhil" (fig. 6).[78]

This "instant" instruction was more rudimentary than what Locke and Milton had in mind when they recommended geography as a suitable subject for the education of young gentlemen. In his essay *Of Education. To Master Samuel Hartlib* [5 June 1644], Milton recommended that young students should read classical authors on agriculture, which would have the double advantage of enabling them thereafter to improve the tillage of their country and of making them masters of an ordinary prose. "So that it will be then seasonable for them to learn in any Modern Author, the use of the Globes, and all the maps first with the old names, and then with the new. . . ."[79] John Locke of Christ Church, Oxford, listed "Arithmetick, Geography, Chronology, History, and Geometry" as subjects to be taught to young gentlemen, preferably in French or Latin, and he proposed Geography and the use of the globe as suitable to begin with, "For the learning of the Figure of the Globe . . . being only an exercise for the Eyes and Memory, a child with pleasure will learn and retain them."[80] He named Heylyn and Moll as the two general English writers in geography suitable for the study of a

77 O. L. Dick, *Aubrey's Brief Lives*, p. xxxiii.
78 Bagford Collection, Harl. 5947, no. 101.
79 John Milton, *Of Education. To Master Hartlib* [1644], p. 4.
80 John Locke, *Some Thoughts concerning Education* (1693), p. 213.

Advertiſement.

GEography made Eaſy, and the Uſe of all the ordinary Sorts of Charts and Maps, whether Geographical, Hydrographical, Plans, Groundplots, or Perſpectives, Taught in a Week's time ; with the Uſe of the Lines in the general Map, and taking of Longitudes and Latitudes, with the Uſe of Scales of Miles ; And an eaſy Explication of the hard Words, which may diſcourage ſome People from this Neceſſary, Pleaſant and Eaſy Science. It is Taught to either Sex, whether Learned in other *Sciences* or not, if they be above the Age of twelve Years. The Maſter Teaches either in his own Chamber, or comes to the Scholars. Price
The Maſter may be heard of at *Mr* Bell's *Bookſeller at the* Bible *and* Croſs-Keys *in* Cornhil.

Fig. 6. "Advertisement. Geography made Easy," with an adjoining pictorial advertisement. British Library, Bagford Collection, Harl. 5947, nos. 100–101.

gentleman, "which is the best of them I know not," adding, "he
cannot well be without Camden's *Britannia* which is much enlarged in the last English edition. A good collection of maps is
also necessary."[81] A less famous author and Oxford man, Edward
Leigh, Master of Arts at Magdalen Hall, produced a useful guide
to travel as the first of *Three Diatribes or Discourses* (1671),
advising the would-be traveler to "inform himself (before he
undertakes his Voyage,) by the best Chorographical and Geographical Map of the Scituation of the country he goes to, both
in it self and Relatively to the Universe, to compare the *Vetus &
Hodierna Regio*, and to carry with him the Republicks of the
Nations to which he goes; and a Map of every Country he in-
tends to travel through."[82] He was also urged that "he should be
first well acquainted with his own Country, before he go abroad;
as to the places and Government"; adding "If any came heretofore to the Lords of the Council for a License to Travel; the old
Lord Treasurer *Burleigh* (William Cecil Lord Burghley, Lord
High Treasurer under Queen Elizabeth I), would first examine
him of *England*; if he found him ignorant, he would bid him stay
at home, and know his own Country first."

Geographical literature was thus prescribed reading for men
of letters and for the schoolroom, where even young girls were
not exempt. A certain Mrs. Charlotte Charke, a lady of some
notoriety in her mature years, recalled in her autobiography
(published in 1755) her childhood lessons. "Nor was my Tutor
satisfied with those Branches of Learning alone, for he got Leave
of my Parents to instruct me in Geography; which by the Bye,
tho' I know it to be a most useful and pleasing Science, I cannot
think it was altogether necessary for a Female. . . . Accordingly
I was furnish'd with proper Books, and two Globes, celestial and
terrestrial, borrow'd of my Mother's own Brother, the late John
Shore, Esq: Sergeant-Trumpet of England; and pored over 'em,
'till I had like to have been as mad as my Uncle, who has given a
most demonstrative Proof of his being so for many Years. . . ."[83]

Neither the lessons at mathematical practitioners' shops, nor
those in the schoolroom, supplied a systematic tuition adequate
for naval men, a matter that caused Pepys deep concern. The
popular instruction in shops and the handbooks on globes, geog-

---

[81] John Locke, *Some Thoughts concerning Reading and Study for a Gentleman* (1751), p. 726.

[82] Edward Leigh, *Three Diatribes or Discourses* (1671), pp. 6–7.

[83] *A Narrative of the Life of Mrs. Charlotte Charke* (London, 1755), p. 26.

raphy, and navigation available for sale belonged to the technical level of Moxon's *Mechanical Exercises,* a world apart (it may now seem) from the achievements in geography, astronomy, and mathematics of Halley or Newton. While the practitioners were expounding "Copernicus for everyman," or, for those who preferred it, the Ptolemaic system, Newton was providing in his *Principia* (1687) the solution to the fundamental problems raised by the theories of Copernicus and Kepler; although Newton himself was a mathematical practitioner who made his own instruments. Professor Taylor describes her book on the mathematical practitioners as "a chronicle of lesser men—teachers, textbook writers, technicians, craftsmen—but for whom great scientists would always remain sterile in their generation."[84] "Mathematicks at that time, with us," recalled Dr. John Wallis, the most famous of the Oxford mathematicians, and Pepys's friend over many years, "were scarce looked upon as Academical Studies, but rather Mechanical; as the business of Traders, Merchants, Seamen, Carpenters, Surveyors of Lands, or the like, and perhaps some Almanack Makers in London. . . . For the Study of Mathematicks was at that time more cultivated in London than in the Universities."[85]

Three official acts in the field of science brought encouragement to those concerned with the improvement of navigation and practical seamanship, and helped to bring the practitioner and the academic together. In 1662, King Charles II founded the Royal Society, giving formal sanction to the "Invisible College" which had been meeting at Gresham College from the time of the Reformation. This "College of virtuosi," as Pepys called it, admitted him to membership on 15 February 1664, and he was elected president in 1684. At its meetings Hooke, Halley, Pepys, and Moxon met on equal terms. The *Philosophical Transactions* [later: *of the Royal Society*], published no less than eighty papers by Halley, including his essay on the trade winds with its accompanying map (for 1686, but not published until 1688) and his two papers on magnetic variation. The society had at once set about establishing itself as a clearinghouse for geographical intelligence. The first volume of the *Transactions* included "Directions for Sea-men, bound for far Voyages," enjoining them to keep "an exact Diary, delivering at their return a fair

[84] E. G. R. Taylor, *Mathematical Practitioners,* p. xi.
[85] Quoted in ibid., p. 4.

Copy thereof to the Lord High Admiral of England, his Royal Highness the Duke of York, and another to Trinity-house, to be perused by the R. Society."[86] In addition to recording observations of physical phenomena, they were "to make Plots and Draughts of prospect of Coasts, Promontories, Islands and Ports, marking the Bearings and Distances, as neer as they can." Charts and coastal profiles were thus seen as basic requirements.

Second, King Charles II founded on 19 August 1673 at Christ's Hospital a Mathematical School, "a Foundation of forty poor boys who having attained to competent skill in the grammar and common arithmetic to the rule of three in other schools in the said Hospital, may be fit to be further educated in a Mathematical School, and there taught and instructed in the art of navigation and the whole science of arithmetic, in order that they may be apprenticed to the sea service."[87] Lorone's state portrait of Charles II, painted for the Hospital, celebrated this event, displaying the king surrounded by objects symbolizing his patronage of navigation, geography, and science. The many trials which beset the mathematical school in its early days were a constant anxiety to Pepys. Elected governor on 1 February 1676, he was to claim that the foundation of the school itself was consummated only "upon Mr Pepys being call^d to ye Secretariship of ye Admiralty; and noe sooner."[88] A volume of papers in Pepys's library relating to the history of the school comprises letters and reports which show Pepys's concern over the failure of the new foundation in carrying out the objects of its foundation. The shortcomings of the teachers both in their mathematical teaching and in their general direction of the boys' education were itemized in detail in his lengthy "Discourse to the Governours of Christ's Hospitall, this 22^{le} of Octr. 1677 touching ye State of the New Royal Foundation there."[89] One major deficiency was remedied when *A New Systeme of the Mathematicks*, written for the school by Sir Jonas Moore, F.R.S., a governor of the Hospital, was published (posthumously) in 1681. It included Cosmography (Part IV), the Doctrine of the Sphere (Part VI), and a New

[86] *Philosophical Transactions*, I (1665/66), 140–142.

[87] J. R. Tanner, ed., *A Descriptive Catalogue of the Naval Manuscripts in the Pepsyian Library*, III (1909), li, n. Also R. Kirk, *Mr. Pepys upon the State of Christ-Hospital* (1935), p.3; and E. H. Pearce, *Annals of Christ's Hospital* (1908), pp. 100–101.

[88] Pepysian MS, quoted by R. Kirk, *Mr. Pepys*, p. 3.

[89] Pepys' Library, Magdalene College, Cambridge, MS 2612, c. 239–267. Quoted by R. Kirk, *Mr. Pepys*, pp. 3–13.

Geography (Part VIII). Sir Jonas's sons-in-law, W. Hanway and J. Potenger, saw the work through the press after Sir Jonas's death, and Edmond Halley checked and revised its geography.

The third major event of Charles II's reign was the founding of the Royal Observatory in 1675, with the express object of "the rectifying the tables of the motions of the heavens, and the places of the fixed stars, so as to find out the much desired longitude of places for the perfecting the art of navigation." In these proceedings Sir Jonas Moore again had played a part. Finding the catalogue of fixed stars so inaccurate, he had proposed privately to build an observatory, and the king translated this private intention into a public act. On 30 June 1675 Moore and Hooke, accompanied by nineteen-year-old Halley, selected a site in Greenwich Park. Once the Mathematical School was established Sir Jonas suggested that the boys should be taken to the observatory to learn the use of instruments.

These three major developments were accompanied by a fourth auspicious event, the appointment of Pepys, first as "Clerk of the Acts of the Navy" (1659–1660), next as Younger Brother of Trinity House (1661–1662) and finally, in 1673, as secretary for the affairs of the Navy. The urgent need for improvements in the seamen's training in navigation, for original surveys of the British coasts, and for improved naval administration were his continual preoccupation. These matters are documented in the state papers of the day and in his own collection of manuscripts and books. His *Naval Minutes* provide interesting sidelights on geographical and cartographical topics. "Recollect what I have learnt from Mr. Moxon, and from Maunsel's old Catalogue of Books, touching the history of our having any globes, maps, or charts in England."[90] "Collect from Mr. Seller a list of sea-charts and books on navigation printed in English, shewing how the books of charts, under several titles, borrow all from Waggener."[91] From 1681 Greenvile Collins, with the encouragement of the king,[92] had been engaged on his survey of the coasts of the British Isles, to be published in *Great Britain's Coasting Pilot* (1693), the first British survey of the British Isles. In his official capacity Pepys expressed misgivings over Collins's claims and competence. "In order to my better judging of Collins's new

[90] Tanner, ed., *Pepys's Naval Minutes*, p. 304. The reference to Maunsel's Catalogue is Andrew Maunsell's *The Catalogue of English Printed Books* (1595).

[91] Tanner, ed., *Pepys's Naval Minutes*, p. 370.

[92] Ibid., pp. 132–133, 144–145.

work, enquire after his sobriety and method of application in performing of the same from some that were with him."[93] "Mr. Hunter [Samuel Hunter of Trinity House] tells me that Trinity House itself complains of Collins's ill performance of his Book of Carts; and yet he dedicates it to them as well as to the King."[94] Pepys had urged Trinity House to take on the survey "but we neglecting, it fell into Collins's hands by virtue of his having served the King."[95] To remedy the omission, Pepys advised Trinity House "that they should make it [the survey] their own by making him [Collins] a Younger Brother, and obliging him by promise to publish nothing of his work without their approval." He further complained of the disregard of his motion, "that he [Collins] should promise not to put the title of our Hydrographer upon anything to be published by him without its being first viewed and approved of by this House";[96] and he deplored "our Trinity House's suffering a work so properly theirs, and in itself of so much moment, to be taken out of their hands and done by a private and single one."[97] The ensign on the Tangier expedition, Thomas Phillips, had similar views: "the only way ever to do such a work well, were to have it done by three several persons, severally at the same time, and a committee of persons of different interests so as to vie with one another, and a committee of persons appointed to sit upon their daily works, and upon comparing them, conclude what is fittest to be taken as good, and where it is needful to have any piece of the work done over again."[98] Such proposals anticipate the foundation of the hydrographic department of the Admiralty more than a hundred years later.

Another piece of original survey discussed in the *Naval Minutes* was the chart of the Strait of Magellan made by Sir John Narborough on his voyage through the Strait of Magellan in 1669–1670. "Mr. Evelyn, from the rudeness of Sir John Narbrough's drawings extant in the Book of Voyages I sent him, observes to me the expectations he has of the effects of our mathematical boys' educations in Christ's Hospital upon that head, and gives me a very proper hint towards illustrating the usefulness

93 Ibid., p. 325.
94 Ibid., p. 388.
95 Ibid., p. 188.
96 Ibid., p. 189.
97 Ibid., p. 301.
98 Edwin Chappell, ed., *The Tangier Papers of Samuel Pepys*, Publications of the Navy Records Society 73 (London, 1935), p. 108.

of drawing in a navigator from the scandalous instances of the want of it visible in Sir John Narbrough's original draught he gave me of the Magellan Streights, and the drawings therein of men and beasts done by his own hand."[99] This criticism plainly is justified by the draughtsmanship of the great chart of the strait made by Narborough now preserved in the Royal Library, which came to the British Museum in King George III's Topographical Collection. Despite the imperfections in draughtsmanship, the printed map dedicated to Pepys, which Thornton published from Narborough's survey, was to rank as the best chart of the strait until the surveys of the 1760s.[100]

Pepys's papers also draw attention to prevailing errors and omissions in navigational practice. Thus in a letter to Pepys of 17 February 1696 on the imperfect knowledge of navigation among seamen, Edmond Halley censors them for their ignorant use of plane charts, and continues:

Another thing they wholly neglect is, to compare & bring together, their Experience of ye Course of ye Tides, so as to reduce them under some general Rule; such as might be useful for those that have occasion to turn to windward. Through ye Channel for instance; Where we find at this present, whole Fleets to lie Wind-bound many Months. . . . And though it must be acknowledged by them, to be a most desirable Piece of Knowledge, and useful in ye Highest Degree; yet 'tis their owne fault that they do not collate their several Experiences, & bring them into a General Synopsis: which would be much for their own use and the Publick service.[101]

[99] Tanner, ed., *Pepys's Naval Minutes*, p. 391. This was probably a reference to Narborough's journal of his voyages in the *Fairfax* and *St. Michael*, 1672 and 1673, now in the Pepys Library (MS 2556). See *Naval Minutes*, 60, 391, note 3; p. 105, note 2. In a letter of 5 November 1700 to Dr. Arthur Charlett, Master of University College, Oxford, Pepys comments on Professor David Gregory's scheme for mathematical teaching, emphasizing the importance of "Perspective," or Drawing, in a young gentleman's education. *Private Correspondence*, vol. 2, pp. 110–111.

[100] Pepys's comment suggests that the large chart entitled "The Land of Patagona" in the British Library, in King George III's Topographical Collection, K.Top.CCCIV.84, may earlier have been in Pepys's possession. A smaller MS chart also survives (Add MS 5414.29). Thornton's printed chart was published in the *Atlas Maritimus . . . . By John Seller* [c. 1673?] British Library, Maps C.8.d.4(29), and there were at least two later issues, in different states. The chart was then reengraved with a dedication to Pepys and published in 1694 in, *An Account of several late Voyages and Discoveries* (1694).

[101] "Papers of Mr. Halley's & Mr. Greave's touching on imperfect knowledge in Navigation &c.," Pepys Library, MS 2185, fol. 6. A copy is in the British Library (Add. MS 30221, fol. 183 et seq).

The search for a general rule from a series of observations brought Halley to the invention of his thematic maps, which were landmarks in the history of cartography.[102] Halley's "general rules" were based for the most part on the observations made on his voyages of investigation which rank as the first truly scientific voyages of discovery of any nation. His achievements illustrate the principle on which Flamsteed the Astronomer Royal discoursed while visiting Pepys on 23 March 1696 "about ye Business of our Math. Sch. & Boys":

I find him to insist very much upon ye Expediency of having a Man of Learning for ye Master there. . . . Besides, (says he), it is most plaine, that all our great attainments in Science, and in ye Chambers & from ye fire-sides (they were his own Words) of Thinking men within doors that were Schollers and Meckanicks, & not from Tarpawlins though of never so great Experience; they generally contenting themselves with making ye best of what they have & doe know, without ever troubling their Heads to look-out for more.

Pepys commented: "I find by him what I always apprehended, that none of our ancient Mathematicians ever apply'd any part of their Study to Navigation. . . ."[103] Halley was preeminent among the "thinking men": "Mr. Hawley—May he not be said to have the most, if not to be the first Englishman (and possibly any other) that had so much, or (it may be) any competent degree (meeting in them) of the science and practice (both) of Navigation? And the inferences to be raised therefrom?"[104]

Among the "tarpaulins" (i.e., the sailors) whom Pepys also knew, the most remarkable was William Dampier, navigator and buccaneer. On 6 August 1698 John Evelyn dined "at Mr. Pepys":

where was Cap: Dampier who had ben a famous Buccaneere, . . . printed a Relation of his very strange adventures, which was very extraordinary, & his observations very profitable: Was now going abroad againe, by the Kings Incouragement, who furnished a ship of 290 Tunn: he seemed a more modest man, than one would imagine, by the relation of the Crue he had sorted with: He brought a map, of his observations of the Course of the winds in the South-Sea, & assured us that the Maps hitherto extant, were all false as to

102 For a discussion of these maps see the final essay in this volume by Norman Thrower, pp. 195–228.
103 Pepys Library, MS 2185, fol.3r.
104 Tanner, ed., *Pepys's Naval Minutes*, p. 420.

the Pacific-sea, which he makes on the S(o)uth of the line, that on the North, & running by the Coasts of Peru, being extremely tempestious.[105]

Dampier's observations on winds, magnetic variation, and other phenomena, though remarkable, lacked the theoretical framework of "the ingenious Author, Capt Hally," as Dampier admitted. "For my part I profess my self unqualified for offering anything of a General Scheme. . . ."[106] Dampier had written nevertheless two of the best travel books in the language. His discoveries are depicted on the maps and pocket globes (c. 1710) of Herman Moll, and were to be fully exploited later by writers such as Jonathan Swift and Daniel Defoe who used geographical intelligence as a major source. Moll features in Gulliver's "Voyage to the Houyhnhnms" as Gulliver's "worthy friend Mr. Moll," who persisted in disregarding Gulliver's correction of the longtitudes of New Holland.[107] Swift's famous quatrain summing up his sceptical attitude to the traditional style of cartography may indeed have been inspired by a chart of this period, Seller's "General Chart of the West India's, 1671," with its striking cartouche (fig. 7):[108]

> So Geographers in *Afric*-maps,
> With Savage-Pictures fill their Gaps;
> And o'er unhabitable Downs
> Place Elephants for want of Towns.[109]

Among the collections of maps and atlases brought together at this time, the most important was the king's private library at Whitehall, where Evelyn on 2 September 1680 saw "Aboundance of Mapps & Sea-Chards,"[110] and where earlier (on 1 November 1660) he had admired "a vast book of Mapps in a Volume of neere 4 yards large,"[111] the atlas presented to King Charles II by Johan Klencke the Amsterdam merchant to celebrate King Charles's accession to the throne and the Restoration of the monarchy. Another notable collection was that assembled

[105] E. S. de Beer, ed., *The Diary of John Evelyn*, V (1955), 295.
[106] William Dampier, *A Voyage to New Holland* (1703), p. 101.
[107] J. Swift, *Travels into several Remote Nations of the World. By Lemuel Gulliver* (1726).
[108] In Seller, *The English Pilot*, Book 4, pt. 4 (1671), plate.
[109] J. Swift, *On Poetry, A Rapsody* (Dublin, 1733), 12.
[110] *Diary of John Evelyn, IV* (1955), 215.
[111] Ibid., III (1955), 260.

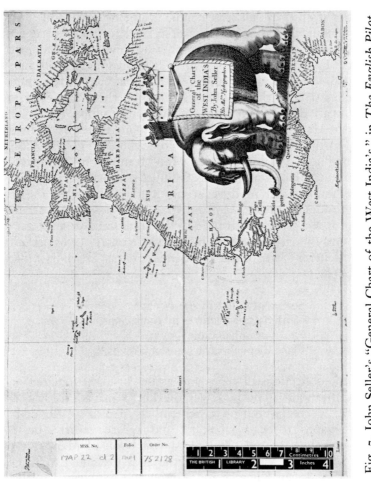

Fig. 7. John Seller's "General Chart of the West India's," in *The English Pilot*, Book 4, pt. 4, 1671. British Library, Map Library, Maps 22.d.2.

by Pepys, recorded in his immaculate catalogue, and now preserved in the Pepys Library at Magdalene College, Cambridge
(fig. 8). The works of Saxton, Waghenaer, Speed, Seller, Moxon,
and Ogilby amongst many others, English and foreign, are well
represented (fig. 9). We can appreciate Pepys's uneasy sleep on
the night of 19 September 1666. "I and the boy to finish and set
up my books and everything else in my house, till 2 in the morning, and then to bed. But mighty troubled, and even in my sleep,
at my missing four or five of my biggest books—Speed's Chronicle—and Maps, and the two parts of the Waggoner, and book
of Cards; which I suppose I have put up with too much care, that
I forgot where they are, for sure they are not stole . . . my books
do heartily trouble me."[112]

No globes are now preserved in the library, no doubt because
globes were too readily discarded when damaged or out of date,
or when more room was needed for books. Perhaps if Pepys had
been able to keep the terrella given him by Edward Barlow his
predecessor as clerk on 2 October 1663, we would find in the
library this little globe—a loadstone acting as a model of the
earth. But "to my great trouble I find it is to present from him
to my Lord Sandwich; but I will make a little use of it first, and
then give it him."[113] This he duly did on 25 November 1663.

It is appropriate to end with the scene depicted in Verrio's
painting (1681–1685), commissioned as a result of Pepys's proposal that King Charles's founding of the Mathematical School
be commemorated (fig. 10). Charles having died at an inopper-
tune moment while the composition was in progress (1685), his
face was painted over with that of the new royal patron. King
James II thus finds himself presiding uneasily over the assembled
ranks.[114] Among officials helping to hold a large map outstretched
is the easily identifiable figure of Mr. Pepys, governor of the
Hospital, master of Trinity House, secretary to the Navy, and
president of the Royal Society.[115]

[112] Latham and Matthews, eds., *Diary* VII (1972), 290. The books were re-
covered on 21 September 1666 (p. 292).

[113] Ibid., IV (1971), 323.

[114] Kirk, *Mr. Pepys*, p. 16.

[115] I am grateful to Mr. N. M. Plumley, Archivist of Christ's Hospital, for
risking life and limb in photographing the detail in Verrio's painting.

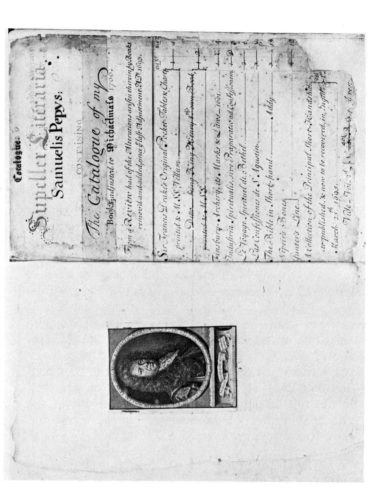

Fig. 8. "Catalogue Supeller Literaria Samuelis Pepys; containing The Catalogue of my Books; adjusted to Michaelmas 1700," fol. 1r; with facing engraved portrait of Pepys after G. Kneller, in Pepys Library. The books and maps listed are all to be found in the library.

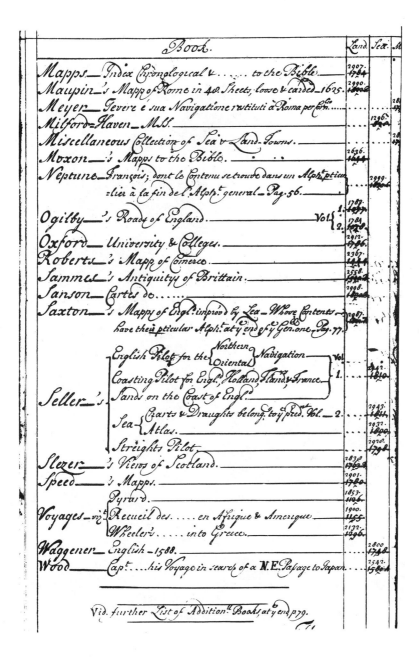

Fig. 9. A page from "The Catalogue to my Books of Geography and Hydrography," by Samuel Pepys. Pepys Library, MS 2700, fol. 601.

Fig. 10. Detail from Antonio Verrio's mural painting at Christ's Hospital, completed in 1685, commemorating the founding of the Mathematical School by King Charles II. Pepys (2nd from the right) is helping to hold with his left hand a large map.

Acknowledgments for Reproductions

Figures 1–7 are published by permission of the British Library; figures 8 and 9 by permission of the Master and Fellows of Magdalene College, Cambridge; figure 10, by permission of Christ's Hospital, Horsham, Sussex.

General acknowledgments

The author also gratefully acknowledges the generous help of Mr. R. C. Latham, Pepys Librarian, Magdalene College, Cambridge.

II

MANUSCRIPT AND PRINTED SEA CHARTS IN SEVENTEENTH-CENTURY LONDON: THE CASE OF THE THAMES SCHOOL

Thomas R. Smith

A study of seventeenth-century sea charts, like most inquiries involving early manuscript or printed materials, depends in large measure upon the resources of institutional libraries.[1] My own concern with the chart trade of seventeenth-century London was strongly influenced by a collection of seventeenth- and eighteenth-century printed maps given to the University of Kansas, for reasons still unexplained, by Otto Vollbehr, who is more widely known for the sale of the Gutenberg Bible to the Library of Congress in the early 1930s. These activities attracted the interest and support of the librarian as well as of the late Joseph Rubenstein, whom Robert Vosper had brought to Kansas to head the Department of Special Collections. I remember visiting the department in response to a seemingly casual invitation to see something interesting. The "something" turned out to be a handsome portolan chart of the Mediterranean, hand colored on vellum, mounted on four wooden panels, signed by Nicholas Comberford of Ratcliff, London, and dated 1666. At that time neither Rubenstein nor myself were greatly concerned with seventeenth-century manuscript nautical charts. But he had purchased it, even though in poor condition, motivated by his own

[1] In his own Clark lecture several years ago, our editor paid graceful tribute to the excellent collections and support for the acquisition of map and atlas materials in the several libraries on this campus. Norman J. W. Thrower, "Edmond Halley and Thematic Geo-Cartography," *The Terraqueous Globe* (William Andrews Clark Memorial Library, University of California, Los Angeles, 1969), p. 3.

bookman's sense for the significance and potential of the item for
a collection where there was scholarly activity.

So it was that I began to investigate Nicholas Comberford and
his contemporaries in what has come to be called the "Thames
School." So it is, with the generous permission of Alexandra
Mason, another Vosper recruit and present head of Special Collections at the University of Kansas, that I have been able to
reproduce the latest of seven charts of the Mediterranean which
Comberford produced between 1647 and 1666 (fig. 1). On it are
the features that characterize the cartography of the Thames
School and relate it to earlier Italian and Mallorcan examples.
It is a plane chart without projection, hand drawn on vellum,
emphasizing the coastline, with names of coastal towns and features at right angles to the shore. The network of radiating
rhumb lines is also characteristic, focusing on a central intersection with the sixteen secondary intersections at a radius from the
center, each with its complement of radiating rhumbs. Thames
School charts of this period were generally decorated with
elaborately drawn compass roses and cartouches around the
scales, embellished with color and gold leaf. Color was also used
on the map itself for print, coastlines, and rhumbs with gold leaf
for some islands.

The lack of detail on land areas leaves room for an index list
of names, keyed by letter or number to the many islands in the
Mediterranean and its included seas. This feature appears unique
to charts of the Mediterranean and has not been found on other
Thames School charts nor on the Italian-Mallorcan forerunners.

Comberford's chart illustrates another characteristic that sets
the Thames School off from most of the earlier portolani. Until
the 1680s they were almost invariably mounted on wooden
panels that were hinged to fold shut with the vellum surfaces
between the boards. This was obviously a protective measure that
facilitated handling and preservation, especially when the charts
were used at sea, but also when housed in shore-based repositories. Despite the widespread practice, the only contemporary
reference found so far is a casual comment by John Seller in *The
Practical Navigator* where, in a brief commentary on English
manuscript sea charts, he noted that they "are drawn on vellum
and pasted on boards and are commonly called Plats."[2]

The Mediterranean chart is still mounted on the original

2 John Seller, *Practical Navigator* . . . , 7th ed. (London, 1694), p. 233.

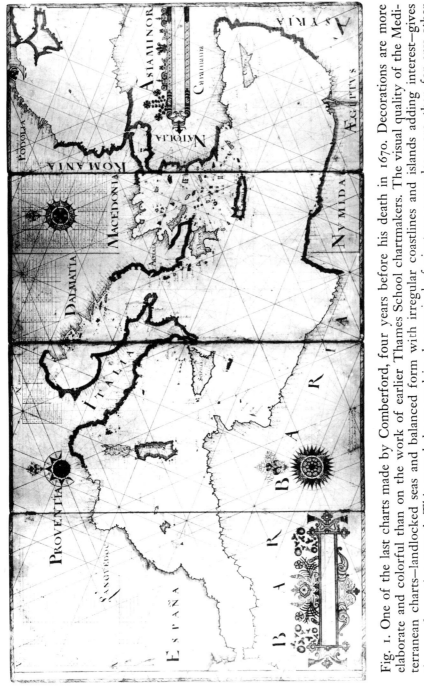

Fig. 1. One of the last charts made by Comberford, four years before his death in 1670. Decorations are more elaborate and colorful than on the work of earlier Thames School chartmakers. The visual quality of the Mediterranean charts—landlocked seas and balanced form with irregular coastlines and islands adding interest—gives them a decorative appeal. This may help to explain the survival of nineteen examples more, than for any other single chart. University of Kansas, Spencer Library, Summerfield MS J7:1.

panels, but they are now completely separate, although remnants of hinges show how they were once fitted together. Four panels were required for the larger oceanic charts (the Mediterranean measures 46½" x 24½" overall) which made it necessary to cut the chart in half, in this case from north to south through the Straits of Messina, so that the center panels could be hinged to fold outward and back on each other. The two outer panels, on the contrary, were attached to the center ones to fold inward with the vellum sharply creased along the hingelines of the two "book folds"—here through the Balearic Islands and the Aegean Sea. Near each end of these creases small oblong slits were cut in the vellum to accommodate the barrels of the metal hinges. When closed, a large chart like this would appear as a stack of four wooden panels with the vellum concealed, and it could be opened full width or by halves depending on need and convenience. Smaller charts could be accommodated on two panels that opened and closed like a book with the vellum intact but sharply creased and with cutouts for the hinges along the center fold.

A number of the surviving Thames School charts are still on original panels, sometimes with hinges intact, perhaps repaired, and operable. An example is Thornton's Indian Ocean, 1682 (fig. 10). More often they have become separated from the panels altogether. But the halving of the vellum and especially the hinge-slits and center-creases form permanent evidence of original board mounting, whatever the subsequent history of the chart. The evidence is clear on the charts themselves and can generally be discerned on the growing number of reproductions in the literature.[3] Also, it should be noted that the division of the

[3] Other reproductions that show the marks of the original board mounting include:

Thomas Hood, Biscay-Channel, 1596, reproduced in David W. Waters, *The Art of Navigation in England in Elizabethan and Early Stuart Times* (London, 1958) opposite p. 200.

Andrew Welch, North Atlantic, 1674, reproduced in Tony Campbell, "The Drapers' Company and Its School of Seventeenth-Century Chart-makers," in Helen Wallis and Sarah Tyacke, eds., *My Head is a Map: A Festschrift for R. V. Tooley* (London, 1973), plate 13.

D. Howse and M. Sanderson, in *The Sea Chart* (New York, 1973) reproduced several Thames School charts from the collections at the National Maritime Museum:

Plate XVI   Thomas Hood, Biscay-Channel, 1956.
      XIX   Burston, Mediterranean, 1638 (detail of parts of the two western panels).
      XXII   Comberford, "South Part of Virginia," 1657.
      XXVII Gascoyne, Caribbean and east coast of North America,

charts for four-panel mounting has resulted in the survival of "half-charts," which present identification problems when the author imprint is on the missing half.[4]

Comberford made this Mediterranean chart (fig. 1) near the end of his career; his last known chart is dated 1670. In the following decades this style was to be virtually abandoned as the Thames School chartmakers continued to produce manuscript charts to meet the changing navigational requirements and shifting area interests of the late seventeenth and early eighteenth centuries. But for the antecedents of Comberford's style and the origins of the Thames School we must turn to the expanding maritime activity in London of the preceding century.

A. H. W. Robinson has written, "the traditions of maritime England, including her marine cartography, go back no farther than the sixteenth century."[5] But after about 1550, change was rapid and London was the scene of an expanding and increasing English maritime activity. The development of the Navy under the Tudors was a partial response to foreign rivalries and threats exemplified by the Armada and the subsequent struggles with the Dutch and the French. Maritime commerce increased as the English economy began to change from the agrarian raw material base. Merchant companies, colonization enterprises, voyages of exploration, even privateering—all increased the demand for British ships and seamen capable and equipped to handle them. The need was great and the development was rapid. David W. Waters in his brilliant and comprehensive study, *The Art of Navigation in Elizabethan and Early Stuart Times*, writes concerning these years:

in 1558 probably not one, as late as 1568 probably only one, English seaman was capable of navigating to the West Indies without the aid of Portuguese, French or Spanish pilots. Yet, by the time of the

---

1678 (on two panels, the west half of a chart of the North Atlantic).

In this paper a good example is plate 7, Daniell, South Atlantic, 1614.

[4] For example, there are two other half-charts of the Caribbean and east coast of North America, similar to Gascoyne's in the list above. They are unsigned but can be attributed to Samuel Welch with an approximate date of 167? on the basis of close identity in line work and print with Welch's chart, North Atlantic, 1674. One of the half-charts is at the Huntington Library (HM 43); the other at the British Museum (Add. MS 31,858). Still another half-chart of the southern part of the South Atlantic is signed by John Thornton, 1681, Library of Congress, Map Division (LC 18).

[5] A. H. W. Robinson, *Marine Cartography in Britain* (Leicester, 1962), p. 15.

Armada, a mere score of years later, Englishmen had gained a "reputation of being above all Western nations, expert and active in all
naval operations, and great sea dogs."[6]

These developments had a major impact on London, whose
trade and importance began to outstrip rival English ports. The
Royal residence at Greenwich and the new naval yards at Deptford, south of the river, were complemented on the north bank
by the growth of commercial docks and supporting maritime
activities. East of the Tower the suburban hamlets grew, spread-
ing houses and streets along the new embankments of the Thames
and filling in the marshes outside the city walls. John Stowe, the
famous chronicler of London, described the changes during his
lifetime. Writing in 1598, he noted the increase in numbers of
houses and shops in the port area of St. Katherine and Wapping
just east of the Tower, where there had been no houses forty
years before. But at the time of writing there was "a continual
street . . . of tenements . . . inhabited by sailors' victuallers, along
by the river Thames, almost to Ratcliffe, a good mile from the
Tower." Of Ratcliff itself, where the first houses had been built
during his youth, he commented "of late years shipwrights and,
for the most part other marine men, have built many large and
strong houses for themselves, and smaller for sailors from Ratcliffe almost to Poplar and so to Blackwall."[7] St. Katherine,
Wapping, Ratcliff, the Minories, Limehouse, and Stepney continued to expand during the seventeenth century, despite repeated prohibitions against building in the London area. The
names of the "Tower Hamlets" are associated with many maritime activities, including chart making. A century after Stowe's
chronicle, Stepney's population had grown to an estimated
75,000. As a result its representation on Gascoyne's map of
Stepney Parish in 1703 is very different from that on the
sixteenth-century maps of London.

It is in these Thames-side parishes north of the river and east
of the Tower that the chartmakers concentrated. The location
is logical because marine cartography was a necessary and integral part of general maritime expansion, especially of the increasing capability and skill in navigation. There was, in fact,
an organized trade in the production and sale of hand-drawn

---

[6] Waters, *The Art of Navigation in England*, p. 101, quoting in part the
Venetian ambassador in France, writing in 1588.
[7] John Stowe, *The Survey of London*, 3d ed. (London, 1618), pp. 792–793.

portolan charts in competition with the emerging output of printed charts. What was the origin of this group of chartmakers and of the style of their work? What was the nature of their output? How were the charts made and how were they used? What justification is there for use of a term "school," and what is the relationship between the hand-drawn platts and the printed charts that ultimately replaced them? Finally, what can be gleaned concerning the chartmaker and his trade in the social economy of the expanding maritime activity of London ports? There are really two inquiries that, at the same time, are separate and inseparable. The charts are studied in an effort to analyze, compare, and evaluate their qualities. The manner of operation, method of work, and actual output of the chartmaker is examined with a sociohistorical perspective. A problem here is that, until recently, most of our knowledge of the chartmakers had been derived from the author imprints on the charts themselves which usually only identify the chartmaker, give his address, and the year of issue.

Serious scholarly interest in these seventeenth-century chartmakers and their work is rather recent, although the decorative quality, especially of the Comberford charts, has given them considerable appeal. Several were exhibited in the British Pavillion at the New York World's Fair in 1939. They also began to appear on the market and at least fifteen have been sold by London dealers since the mid-1920s. Coming from English private collections, they are all now in public or university libraries: half are at the National Maritime Museum in Greenwich, the remander being at Harvard, Yale, Dartmouth, and Kansas.

Important contributions to the literature in this field have been cartobibliographic in character, listing manuscript charts and generally concentrating on continental collections and chartmakers. Information on the English counterparts has been meager and incidental. In 1941 Marcel Destombes, the French authority, pioneered in a survey of manuscript nautical charts produced by authority of the Dutch East India Company.[8] Destombes has continued his interest and broadened his perspective to include chartmakers outside of Holland. A monograph jointly authored by Pastor and Camarero, *La cartografia Mallorquina*, published in 1960, included a list of the portolani of Mediterranean origin

---

[8] Marcel Destombes, *Catalogue des Cartes Nautiques Manuscrites sur parchemin, 1300–1700, cartes Hollandaises* (Saigon, 1941).

in collections in England and on the continent.[9] A third and more recent publication of this type is the catalog of nautical charts on vellum at the Bibliothèque nationale in Paris, authored by Mlles Foncin, de La Roncière, and M. Destombes[10] and based on a large collection, including charts by English as well as continental cartographers.

It is natural that the London chartmakers should receive more attention from English scholars. In the major works of Robinson and Waters, which I have already noted, we find brief mention of only three or four of the seventeen Thames chartmakers who have thus far been identified. More attention is paid to these men by the late E. G. R. Taylor, professor of Geography at the University of London. In *Mathematical Practitioners of Tudor and Stuart England,* she included brief biographical information for five members of the group: Daniell, Burston, Comberford, Gascoyne, and Hack, but like her two compatriots, she restricted her comments to information derived from the charts. However, this did not preclude an important suggestion in her note on Burston where she wrote, "The similarity of his address—at the Sign of the Platt in Ratcliff Highway near London—to that of Nicholas Comberford as well as the similarity of their work suggests a partnership, although it may have been a rivalry."[11]

It would appear that Professor Taylor's suggestion of Thamesside cartographic collaboration bore first fruit in Spain where, in 1959, Camarero published a short article whose title translates as "The English School of Cartography 'At the Signe of the Platt.'"[12] In the process of canvassing European libraries for charts by Spanish and Italian cartographers, Camarero had seen some of the English charts and was familiar with Professor Taylor's book. His article lists about thirty charts by Burston, Comberford, John Thornton, Welch, and Gascoyne—with incidental reference to Daniell and Hack. Camarero also commented on the similarity of the work of these men and pointed to its derivation from the Mediterranean cartography. He noted also the frequent use of the designation "at the sign of the Platt" in the author

[9] J. R. Pastor and E. G. Camarero, *La Cartografia Mallorquina* (Madrid, 1960).

[10] Myriem Foncin, Marcel Destombes, and Monique de La Roncière, *Catalogue des cartes nautiques . . .* , Bibliothèque nationale (Paris, 1963).

[11] E. G. R. Taylor, *Mathematical Practitioners of Tudor and Stuart England 1485–1714* (Cambridge, 1954), p. 245; also pp. 202, 228, 270, 275.

[12] E. G. Camarero, "La Escuela Cartografica Inglesa 'At the signe of the Platt,'" *Publicaciones de la real sociedad geografica,* Serie B., Numero 399 (Madrid, 1959).

imprints, and seems to have interpreted the phrase as an indication of collaboration at a single address. But from what is known about individual chartmakers, it is clear that "signs of the platt" were at different addresses and several have been located by Professor Taylor.[13] The term refers to a board with a chart or platt painted on it and hung outside the shop as a means of identifying the premises.

So far, then, the published material was sketchey, scattered, peripheral, and based entirely on the signatures and a casual comparison of some of the charts. However, at a local library in London a volume was found, privately printed in 1890–1901, which contained records of St. Dunstan's, the parish church of Stepney. Among these was a notice of Comberford's wedding in 1624 which identifies him as "Nicholas Comberford, of the precincts of St. Katherine, draper."[14] The word "draper" is the key. The possibility that Comberford had been a member of the Worshipful Company had never occurred to me nor, so far as I am aware, to anyone else. A check of the published *Roll of the Drapers Company* by Percival Boyd verified Comberford's membership but added nothing else, since the volume consists only of a name list with pertinent dates.[15] In Johnson's four-volume history of the company, among those members assessed for the poll tax in 1641, three "plattmakers" are listed: Comberford, Daniell, and Solmes, who were reported to have a total of fourteen apprentices, although none were named by Johnson.[16]

This find constituted the second primary source on the Thames

13 E. G. R. Taylor, *Mathematical Practitioners*, map "The Work Places of the London Practitioners," opposite p. 162.

14 G. W. Hill and W. H. Frere, eds., *Memorials of Stephney Parish . . . 1579–1662* (Guildford, 1890–1891), pp. 183–189. That my own initial efforts were concerned with Comberford was fortuitous, as I have explained. It was also fortunate in several important ways, one being that Comberford led to a chance discovery by a colleague whom I had asked to investigate this matter of an entirely unexpected source documenting a series of master-apprentice relationships and connecting major chartmakers in sequence for more than a century.

15 Percival Boyd, *Roll of the Drapers' Company of London* (London, 1934).

16 A. H. Johnson, *The History of the Worshipful Company of Drapers* (Oxford, 1914, 1922), IV, 142, 152, 158.

Unable to visit London at this time, I arranged for a professional researcher to make preliminary checks at the Drapers' Hall. It was not until the summer of 1969 that I had the opportunity of examining the records in person and found much additional information. The most available source is Boyd's unpublished manuscript index, transcribed from the original records and arranged alphabetically, which links masters and apprentices by name and generally gives dates, addresses, the trade or craft practiced by the individuals, as well as information on the payment of quarterage and other individual relations with the company.

chartmakers, independent of the author imprints on the charts and providing more information. It is from these records at the Drapers' Hall that the diagram of master-apprentice relationships has been derived (Appendix 1).[17] The master-apprentice tree includes names of thirty-seven individuals, with twelve of them identified as having produced at least one signed chart that has survived. Seven of the dozen starred individuals are arranged in a direct master-apprentice sequence of seven "generations" which forms the central trunk of the tree. At the top is the unstarred James Walsh about whom the Drapers' records tell us little other than that he granted the freedom to P. Walsh and to John Daniell, the first chartmaker in the sequence. Daniell was followed (in time sequence and downwards on the diagram) by Nicholas Comberford, John Burston, John Thornton, Joel Gascoyne, and finally the Friends, John, the father, and Robert, his son.

The twenty-five unstarred names represent individuals who appear in the apprenticeship records, but for whom no signed charts have been found. Note that most of them did not complete their apprenticeships since only a starting date is shown. Perhaps this represents wastage in the system. But several of these men were listed in the records as plattmakers or sea card drawers. Solmes was one of the three identified as plattmakers by Johnson, and the diagram shows that he completed his apprenticeship and himself had apprentices. The same is true for Thomas Comberford and Wild, so there must have been cartographic activity which may be represented in anonymous labor on charts signed by others, in signed charts yet to be discovered, or in the considerable number of anonymous charts that have been preserved.

[17] At the Third International Conference on the History of Cartography at Brussels, September 18, 1969, an earlier and less complete version of this master-apprentice tree was displayed and discussed in my paper, "Nicholas Comberford and the Thames School of Sea-chart Makers of Seventeenth Century London." A revision of that paper, including master-apprentice tree in nearly the present form has remained unpublished but was distributed in mimeographed form to interested scholars in October–November 1969. Following my initial investigation of the Drapers' connection, others have worked with the materials at the Drapers' Hall, notably Mr. Tony Campbell and Professor Coolie Verner, in part in connection with the latter's research on John Thornton. In the period since my Clark Library lecture was given 4 March 1972, Campbell has published a master-apprentice tree consisting of the thirty-seven names. This indicates only that at least two investigators derive similar data from the Drapers' records: see Tony Campbell, "The Drapers' Company Chart-makers," in Wallis and Tyacke, eds., *My Head is a Map*, pp. 81–106, especially the Apprentice Tree, p. 100 and Biographical Table, p. 101.

The master-apprentice record spans approximately a century and a quarter and documents close working relationships that go beyond the convenient proximity of work places. The connections proposed here are doubtless incomplete, and there may be interactions between the individuals in the diagram which cannot yet be documented. But those connections and interactions which are demonstrable reveal the basis for transmission of skill, technique, and style as well as the transfer of actual cartographic material which may help to explain the similarity in the charts produced by various cartographers for over a century. Here, too, in this mechanism of instruction, is a more solid basis for the suggestion of a "school" than in the similarity of address suggested by "the sign of the platt." In casting about for a name, one is tempted to suggest the "Drapers' School," but this is hardly satisfactory for chartmakers. It is Jeannette Black's term, "Thames School,"[18] which has come to be used to designate the group of seventeenth-century London chartmakers who produced hand-drawn charts of portolan type and whose places of work were concentrated along the river which was the natural focus for ships, mariners, merchants, and the school's emphasis on sea charts.[19] Although the Drapers' Company formed an organizational framework, the Thames, and its port, provided the raison d'être for the chartmakers and therefore takes precedence in the designation.[20]

Why should chartmakers have been members of the Drapers' Company? There is no definite answer because the records are sketchy and incomplete. Since so little is known about Walsh, there is no clear explanation for Daniell's apprenticeship. Was Walsh a chartmaker? Was the apprenticeship a matter of convenience? Perhaps it had nothing to do with chart making, and Daniell may have developed the skill in the period between en-

[18] The priority of Miss Black in this regard was recognized in my paper at the Brussels Conference in 1969. In the spring of 1975 *The Blathwayt Atlas*, Volume 2: *Commentary*, has appeared to complete the work initiated by the appearance of Volume 1: *The Maps*, published in 1970 by the Brown University Press. In her *Commentary*, Miss Black devotes an introductory chapter to "The Manuscript Maps and the Thames School" (pp. 15–22), with additional comment in her detailed discussion of the individual maps.

[19] For example, see Howse and Sanderson, *The Sea Chart*, especially the discussion of plates XIX, XXII, and XXVII.

[20] Campbell, as is obvious from the title of his paper, favors the alternative choice ("The Drapers' Company Chart-makers," pp. 98–99), but he does not treat the sixteenth-century origins of the style by non-Draper chartmakers (see pp. 57–65).

tering the freedom in 1590 and his first known chart dated 1612. This leaves us with the consideration of possibilities and inferences.

The possibility was present because there was nothing to prevent a chartmaker from becoming a Draper. By the late sixteenth century the craft monopoly and influence of the guilds were rapidly weakening. Referring to this period, Johnson points to the large "number of Drapers who are no longer pursuing the mystery of Drapery at all."[21] Apprenticeship and freedom of the Drapers Company was available to the followers of many trades and crafts as the various lists in Johnson clearly indicate.

More positive influences may have favored the inclusion of the chartmakers in the Worshipful Company. During Elizabeth's reign, for example, Drapers became increasingly involved with the activities of the newly formed merchant companies. Some Drapers were known to have been members of such companies and may have participated either as merchants or as seamen in voyages of trade or discovery. Johnson quoted Drapers' records of the 1560s listing cloth and other merchandise, presumably owned by Drapers, and laden on ships engaged in the growing trade with Russia. In the Southern Trade, Drapers were also involved in early trading voyages in Africa.[22] From these indications, it can be inferred that there were few barriers to membership and various circumstances which increased such a possibility. Since chartmaking was a developing craft in Elizabethan England —too small and too late to form its own guild—it was no doubt in need of a master-apprentice mechanism. But this "need" was not necessarily a motivating force. More likely the development of the Drapers' connection was a chance occurrence with the probabilities favoring a large and powerful company with wide mercantile and marine interests but with no craft restrictions.[23]

[21] Johnson, *The History of the Worshipful Company*, II, 163.
[22] Ibid., pp. 177 ff., 456–460.
[23] In regard to this point and the alternatives suggested in the preceding paragraphs, Campbell has found strong circumstantial evidence that James Walsh (Welsh?), who heads the master-apprentice tree, was a seaman and that Daniell was apprenticed to him in this capacity. Boyd's transcript lists Daniell as a compassmaker and later (?) as a seacard-drawer. The occupational sequence appears to have been from seaman to compassmaker to chartmaker, probably with overlap. Such a transition is logical, not unusual, and in this instance establishes a probable connection between the chartmakers and the maritime activities of the Drapers. At least one other individual named on the tree, Charles Wild, was also a seaman, and presumably his apprentices were as well. Campbell lists several map and chartmakers of the period who were

If we turn from the chartmakers to the charts themselves, we can trace the origins of the Thames School style to the last decade of the sixteenth century and the work of several chartmakers, of whom there is no record at Drapers' Hall. The style was well developed in the charts, first of Thomas Hood and Gabriell Tatton and later in the work of Nicholas Reynolds and Thomas Lupo. The emphasis was on navigation charts for large areas of ocean, similar in design and style. A standard area coverage was developed for the Mediterranean and the North Atlantic by about 1600 and later for the South Atlantic, Indian Ocean, and China Sea. Similarities among the very early charts imply that the chartmakers worked together or at least influenced each other. But there is as yet no master-apprentice tree to document this interaction.

Thames School charts of the Mediterranean offer a point of beginning for illustration and elaboration of these generalizations. This sea was a favorite of the London chartmakers of the late sixteenth and seventeenth centuries, and during this time members of the Thames School produced nineteen known charts of the Mediterranean—all of them similar, some nearly identical. Seven of these were made by Comberford, another half dozen by Burston, two by John Daniell, and the remaining four by as many chartmakers (Appendix 2). The first in the series is by Gabriell Tatton who signed himself with the address "att the Signe of the Goulden Gunn att the Weste ende of Ratclyff" (fig. 2). The date, 1506, above the author imprint is obviously wrong and probably a forgery since there are signs of erasure at the spot below the imprint, which is the usual location for dates on these manuscript charts. An approximate date of 1600 is consistent with what little is known of Tatton's other work.[24]

---

members of other livery companies ("The Drapers' Company Chart-makers," pp. 88–93).

One of the latter was John Seller who made several manuscript charts on vellum in Thames School style. Note that there were Draper masters / apprentices who were not chartmakers, as well as practitioners of the latter trade who were not Drapers.

24 The chart is at the Newberry Library, Ayer MS Map #22. The date, 1506, appears to have been altered and the third digit has been variously read. It is in a different hand from the print on the chart which further supports the hypothesis that the original date has been erased from the more usual location below the author imprint.

This chart was sold to Grossmith at Sotheby's sale of 12 April 1899, Item 578, and the catalog gives the date uncertainly as "1506 (1566)." But this does not inhibit the provocative note "A highly interesting and valuable Portolano: said

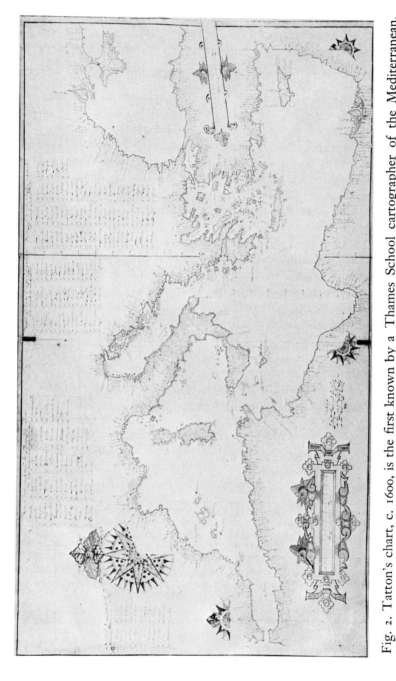

Fig. 2. Tatton's chart, c. 1600, is the first known by a Thames School cartographer of the Mediterranean. The given date, 1506, is incorrect as explained in text. The general design and coastal alignment is remarkably like the Comberford chart of 1666, although decorations are less elaborate and there is only a half circle of compass roses and rhumb intersections. Chicago, Newberry Library, Ayer MS Map 22.

Tatton's chart is similar in design and identical in area to that of Comberford of 1666; comparison of the two reveals that the main elements of the Thames School format were already present in the Tatton chart: portolan style, rhumb network, compass roses and cartouches, the arrangement of graphic and latitude scales, and the lists of island names. Color and gold are used on both, but less by Tatton whose style and decorations are more restrained than Comberford's. Also the Tatton chart was once mounted on boards and, being smaller than the later ones, it could be accommodated on two panels as indicated by the center crease and hinge-slits at the top and bottom. Tatton used a scale of about 23 leagues to the inch (as measured from the graphical scale), while a larger scale of 15 leagues to the inch became standard on Thames School charts of the Mediterranean beginning in the 1640s. Finally, it can be observed, even on a reduced reproduction, that the chart is well drafted with regard to line work, printing, and decorations.

Two other charts of the Mediterranean were produced in the late sixteenth century or very early years of the seventeenth century. One by Nicholas Reynolds is still on twin panels and quite similar to the Tatton chart but with less decoration.[25] In the British Museum[26] is a third Mediterranean chart of the period by Thomas Lupo, whom we know only from his signature on the chart, which is undated but assumed to be c. 1600. A striking feature of the three is that the line work for the coasts is so close

---

to be the only English one known of this early date" (sic!).

Alterations or forgeries of dates are not unknown, especially on manuscript maps and charts. In fact Tatton's chart of Guiana (British Museum, Add. MS 34,240N) bears the implausible date of 1688 in the proper place on its face. Authorities differ, thus a Mr. Coote, one-time curator of maps at the British Museum, dates the map 1608, while C. A. Harris considers a date around 1628 more likely. The latter suggestion has been disproved by research in progress since this paper was presented. Mrs. Sarah Tyacke, assistant keeper, Map Room, British Library has been engaged in a detailed study of Tatton. She has found definite record of his death in 1621 and believes a more likely date for his chart of Guiana is c. 1613. The Mediterranean chart herein discussed is the first known example of his work which now comprises eight charts some of which were previously unidentified, as well as an atlas of seventeen charts of the East Indies. Mrs. Tyacke considers Tatton one of the leading London chartmakers of his time (personal communications). Those interested in seventeenth-century hydrography look forward to the publication of her findings. See C. A. Harris, *A Relation of a Voyage to Guiana by Robert Harcourt*, Hakluyt Society, ser. 2, 60 (London, 1926), 17–18.

[25] Florence, Biblioteca Nazionale, Port. #11.
[26] British Museum, Add. MS 10,041.

as to have been traced. This similarity was determined by super-imposing images in an enlarging-reducing projector which re-vealed an usually close concidence in representation of the coasts of the three charts which are alike in size and nearly identical in scale. Likewise the area covered by all three charts is very similar, except that Reynolds shows all of the Black Sea by means of a "fold out" flap of vellum attached to the upper edge of the right-hand panel, which extends the chart to the east. But the main rectangle covers the entire Mediterranean from just west of Gibraltar to the western half of the Black Sea. These charts by Tatton, Reynolds, and Lupo established the standard design and area for the Thames School representation of the Mediterranean which was to continue for over eighty years. The close similarity of the charts indicates mutual awareness or a common source, if not actual collaboration between the three chartmakers.

The North Atlantic was the other water body for which a standard delineation was developed during this early period. We currently know of fourteen charts of the North Atlantic by Thames School chartmakers dating from 1592 to 1678 (Appen-dix 2). Thomas Hood, the fourth of the pre-Draper chartmakers, is better known as a mathematician and lecturer on navigation.[27] But he produced several important charts, both printed and manuscript, between 1592 and 1604. His chart of the western Atlantic, signed "Thomas Hood / made this Platte / 1592," is preserved in the Staatsbibliothek in Munich among the manu-script charts for Dudley's *Arcano del Mare*.[28] It is known to me only from a photograph reproduced as figure 3 which reveals it

[27] Hood and Seller are the only Thames School chartmakers to be recognized in the *Dictionary of National Biography*, and for reasons not connected with manuscript charts. Hood's note of nearly a page (IX, 1164) is devoted to his activity and publications in the field of mathematics and navigation. Seller has a somewhat shorter note (XVII, 1165–1166) noting his post as Hydrographer to the king and his publication of the atlas *The English Pilot* and several books.

[28] Staatsbibliothek, Cod. icon. 140 no. 84. Four volumes of manuscript charts for the *Arcano del Mare* are preserved in Munich. The Hood chart is apparently one from the reference collection which Dudley is thought to have used. The manuscript text for the *Arcano* is preserved in Florence where Dudley lived until his death in 1649, an exile from England. There, in the Biblioteca Nazionale, is the largest collection of early Thames School charts outside of England, thought to be a part of Dudley's original collection. This and related matters receive considerable attention from I. N. P. Stokes in *Iconography of Manhattan Island 1498–1909* (New York, 1915–1928), 11, xxvi–xxix. Stokes depends on F. C. Wieder who conducted the inquiries for him in 1912 and 1913 and who himself published on this complex problem; "Onderzook naar de oudste kaarten van de omgeving van New York," *Tijdschrift van het Koninklijk Nederlandsch aardrijkskundig genoutschap* 35 (1918), 235–260.

Fig. 3. Signed by Hood and dated 1592, this is the earliest known chart of the Thames School. It shows Hood's characteristic restraint in decorations and careful drafting as well as the main elements of Thames School style, except that it appears never to have been mounted on panels. The chart once belonged to Robert Dudley and a vessel track is visible from south to north, right of center, which may have been of his voyage to Guiana, 1594. Munich, Staatsbibliothek, Cod. icon, 140, Nr 84.

to be carefully drawn on a single sheet of vellum with no discernible indications of previous mounting. It is a complete chart with a border on all four sides and a full complement of rhumb lines radiating from the center between Cuba and Hispaniola to the sixteen secondary intersections that fill the frame. Line work and printing are neatly executed as are the rather restrained cartouches and compass roses some of which, unfortunately, have been cut out. This is the earliest signed and dated Thames School chart and an excellent example of the style. It is also very similar in style and design to Hood's chart Biscay-Channel, 1596 which is more widely known from reproductions that show it to have once been mounted on boards in true Thames fashion (see note 3).

These signed charts help in the identification of an unsigned, undated, unmounted chart of the North Atlantic (fig. 4) as probably the work of Hood and the earliest Thames School chart of the area. It is drawn to the same scale as Hood's chart of the western Atlantic, about 45 English leagues to the inch, so that comparison is easy in the areas of the New World common to both. Here the coastal delineation is nearly identical and there is close similarity in handwriting, style, and design. The compass roses, for example, have a patterned design like those on Hood's signed chart. These various similarities support an apparent attribution by Commander Waters[29] and allow the chart to be more accurately identified as (Thomas Hood) North Atlantic (159–?).

A few years later another manuscript chart of the North Atlantic was drawn and signed "Gabriell Tatton made this Platte / in London Ann. Dm / 1602."[30] Tatton's chart (fig. 5) is on the same scale as Hood's, similar in design and style, but with slightly more elaborate decorations and obviously by a different hand. It also covers the same area from about 55°N to 5°S latitude, with the British Isles and the Channel in the northeast corner and the coasts of Europe and Africa to the Gulf of Guinea. On the west the two charts show the coast of the Americas from the Amazon to Newfoundland including the Caribbean and all of the Gulf of Mexico. Suitable for navigation between western Europe and anywhere north of the equator in the New World, this became

[29] The chart is at the British Museum, Add. MS 17938B. Waters apparently refers to it in a note citing "a world (sic) chart by Thomas Hood, BM. Add. MS 17,938 (sic) . . ." somewhat inaccurately in regard to area and number, p. 229n.

[30] Florence, Biblioteca Nazionale, Port. 21.

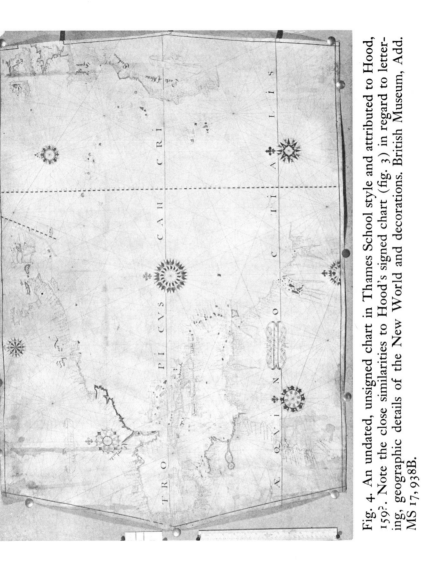

Fig. 4. An undated, unsigned chart in Thames School style and attributed to Hood, 159?. Note the close similarities to Hood's signed chart (fig. 3) in regard to lettering, geographic details of the New World and decorations. British Museum, Add. MS 17,938B.

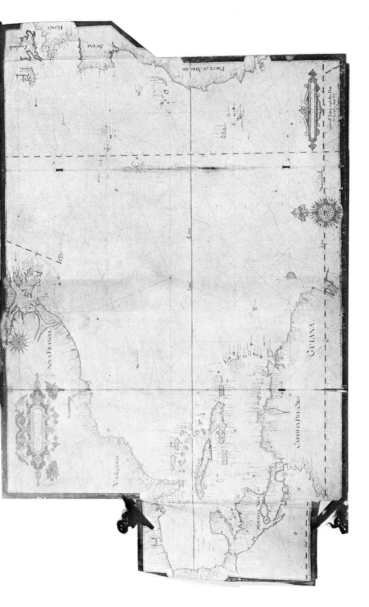

Fig. 5. Tatton's chart of the North Atlantic, 1602, is as similar to the one attributed to Hood as are the later charts of the Drapers, like Comberford, to those of Burston and Daniell. Elements of Thames School style include mounting on wood, although with an unusual combination of one wide and two narrow panels (note hinges). "Fold in flaps" extend the chart area on each side beyond the edges of the panels, a device noted on one or two other contemporary charts. Florence, Biblioteca Nazionale, Port. 21.

the standard Thames School chart of the North Atlantic for the series which ended with Gascoyne's half-chart of 1678 which has been recently reproduced in Howse and Sanderson.[31]

With a cartographic style well developed by 1600, there can be no doubt that the origins of the Thames School are rooted in the expansion and development of maritime activities during the Elizabethan period, but specific connections and prototypes are as difficult to establish as is the explanation for the later platt-makers as Drapers. Robinson has pointed out that most of the English sea charts of the sixteenth century were harbor plans or pilotage charts for ports and other restricted areas. However, navigation charts for oceans and seas were beginning to be pro-duced by seamen like William Borough as a result of longer voyages for discovery or trade. The brothers William and Stephen Borough were renowned pilot-captains and agents for the Russia Company and later served the queen with distinction in various capacities. Robinson considers William to be the fore-most nautical cartographer under the Tudors[32]—approximately ten charts attributed to him are known, including several por-tolans. An example is Borough's chart of the North and Baltic Seas (fig. 6) undated, but thought to be c. 1580. Howse and Sanderson comment: "Though the chart looks unfinished it was probably designed specifically to show only those coasts that would be seen by a ship on passage from England to Narva in the Gulf of Finland. . . ."[33] The shape of the vellum and the layout and form of the compass roses are similar to the North Atlantic chart attributed to Hood (fig. 4). Likewise, Hood's manuscript chart of the coasts of England, reproduced in Skelton and Summerson,[34] shows similarity in draftsmanship to the works of Borough. There is also a considerable similarity between Hood's manuscript Biscay-Channel, 1596 and Edward Wright's charts of the Northeast Atlantic, especially the manuscript ver-sion of 1599 preserved at Hatfield House.[35]

From this it is clear that the two decades bracketed by Hood's chart Western Atlantic, 1592 and Reynolds's Mediterranean of c. 1612 mark the emergence of the Thames School. Already pres-

[31] Howse and Sanderson, *The Sea Chart*, Plate XXVII, p. 69 (see n. 3).

[32] Robinson, *Marine Cartography*, introduction and especially pp. 29 ff.

[33] *The Sea Chart*, p. 39 and plate XII.

[34] R. A. Skelton, and John Summerson, *A Description of Maps and Archi-tectural Drawings in the Collection Made by William Cecil, First Baron Burgh-ley, Now at Hatfield House* (Oxford, 1971), plate XV and p. 39.

[35] Ibid., plate XIII and pp. 70–71.

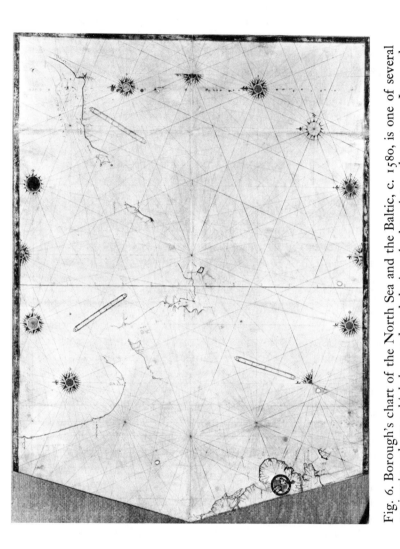

Fig. 6. Borough's chart of the North Sea and the Baltic, c. 1580, is one of several navigation charts which he produced during the late sixteenth century. It reveals some similarities to the Thames School but can hardly be regarded as a prototype for the style that emerges with apparent suddenness in the charts of Hood and Tatton, 1592 and later. Greenwich, National Maritime Museum, G215: 1/5.

ent were the attributes that characterized its later products—adherence to the traditional portolan form, a consistent cartographic style including line work, use of color and gold, cartouche and wind-rose decorations, and island-lists on the Mediterranean charts. Hood, Tatton, Reynolds, and probably Lupo are the first Thames School hydrographers in the sense that they produced similar charts in a style and form which were to be followed and elaborated by others for nearly a century. The charts were well drawn and carefully laid out, indicating a rather advanced state of cartographic practice, especially in the case of Hood and Tatton. The similarities that have been noted support the conclusion that there must have been interaction and perhaps collaboration among them. Skelton has surmised:

In the last years of the sixteenth century, shops for the production, copying and sale of manuscript nautical charts seem to have similarly operated in the Thames-side parishes frequented by seamen. Robert Norman the compass-maker signed his only two surveying charts (both collected by Burghley) [at Hatfield] from Ratcliffe; Thomas Hood, whose extant signed charts are dated from 1592–1604, had a shop in the Minories; Gabriel Tatton, from c. 1586–1612, also worked in Ratcliffe; and John Daniell probably from 1590, in St. Katherine's by the Tower.[36]

If there were workshops there must have been apprentices and masters, but no documentary evidence has yet been found and the names of the four earliest Thames chartmakers do not appear in the Drapers' records.[37] Even though they were contemporaries of John Daniell, the first Draper chartmaker, who entered into the fredom of the company in 1590, there seems to have been no close connection. Yet Daniell's style is clearly related to that of Hood and Tatton.

The earliest chart signed by Daniell is of the South Atlantic (fig. 7) with the imprint in a modest box beneath the scale, bottom left, "Made by John Daniell in St. Katherine's nere unto the Tower of London. Anno. 1614."[38] The chart is quite similar to those of Hood and Tatton which have been discussed (figs. 2–5); the style is simple, not flamboyant, and with modest com-

---

[36] Ibid., p. 25.

[37] Error is doubtful in view of two careful and independent searches of these records by Campbell and myself.

[38] South Atlantic, 1614, British Museum, Add. MS 1514.C.r.

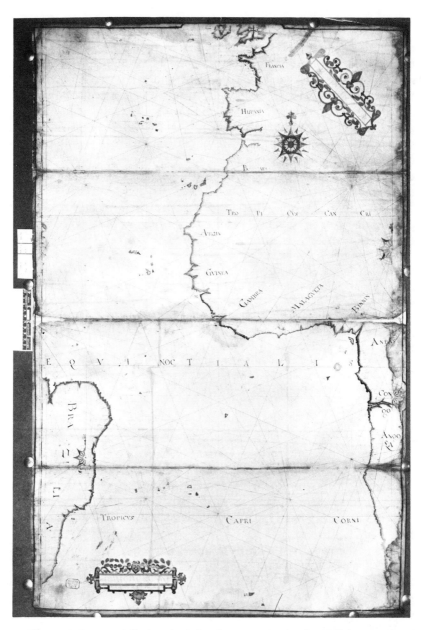

Fig. 7. John Daniell, South Atlantic, 1614, is the first Thames School chart of this important ocean and the first of record by a member of the Drapers' Company. Contemporary references have been found to at least two other charts by Daniell 1614 or earlier and this one is similar in style and content to those of Tatton and Hood. British Museum, Add. MS 5415.C.1.

pass roses and scale decorations. The use of color is restrained, the drafting carefully done and the lettering neat and fine.

Tatton's South Atlantic chart also established a standard representation that was closely copied on five later Thames School charts, the series ending with Thornton's of 1681. On the east side are the coasts of Europe and Africa from the English Channel to the Cape of Good Hope, but with the Mediterranean omitted. Opposite is the bulge of Brazil from the mouth of the Amazon to the River Plate, while some of the charts show the tip of Newfoundland in the northwest corner. The six are all large charts, mounted similarly on four horizontal panels to accommodate the greater north-south dimension. Rhumb networks are also similar to that on the Daniell chart, with the central intersection close to the African coast about $7\frac{1}{2}°N$. The scales are all very close to 45 English leagues to the inch as measured from the graphical scales. There are differences in coastal alignment and nomenclature, as is to be expected in a series spanning some seventy years. But the standard area and general design were followed very strictly as in the case of the Thames School charts of the Mediterranean and the North Atlantic. This representation of the South Atlantic was obviously useful for voyages to Brazil and the west coast of Africa. But it seems probable that its chief use would have been on runs between English ports and the Cape on the route to and from India and the East Indies. This traffic was of growing importance during the seventeenth century and charts for this Oriental Navigation became a major concern of the Thames School hydrographers during its closing decades. Before turning to this development, however, the seas closer to "home waters" remain to be considered.

It is only natural that the seas adjacent to western Europe should have engaged the attention of the Thames School chartmakers. There are sixteen known charts that can be grouped in a loose category "western Europe," with eleven signed by either Comberford or Burston and the remainder by Hood, Daniell, or John Thornton. They are also concentrated as to period, with ten of the sixteen dated in the 1660s and only one later than this decade. Whether this "timing" represents an emphasis of productive effort or the accidents of survival is impossible to determine (see Appendix 2).

The earliest is the fourth known manuscript chart by Thomas Hood, Biscay-Channel, 1596, which has been reproduced at least twice in the literature.[39] These reproductions reveal the neat and careful draftsmanship that characterizes Hood's work and also, unlike his other manuscript charts, it is mounted on two wooden panels in true Thames style. Also it appears as the prototype for five later charts Biscay-Channel produced by Comberford or Burston between 1641 and 1668, of which Burston's of 1664 is reproduced here (fig. 8). Burston follows Hood in showing a large number of soundings in the Channel and off the coast of France as far as the one hundred fathom line—an unusual feature on Thames School navigational charts. But he does not include the data on the tidal establishment of the ports which was featured by Hood. These generalities also apply to other Biscay Channel charts, but they differ from each other in area and coastal delineation. For instance, Burston's chart of 1664 shows more of the British Isles than does Hood's, while Comberford's 1666 chart is a large four-panel map extending north to the Shetlands and southern Norway. The two remaining charts (Daniell, 1626 and Burston, 1660) focus on the Channel but extend west to the Azores and south to the Canaries. In this they are similar to an engraved chart by Hood of the Northeast Atlantic, 1592.[40]

The Northern Navigation also engaged the attention of the Thames School, and three of the four surviving charts for this purpose are quite similar. One by Comberford 1665 (fig. 9) is a four-panel chart with the British Isles at the bottom extending north to Iceland, Greenland (Groinland), and Spitzbergen (Greeneland), and including all of Scandinavia, the Baltic, and the arctic coast of western Russia. Two others are quite similar while the fourth, by Comberford in 1665, is of the North Sea and the Baltic, an area similar to that covered by Borough's chart (fig. 6) of nearly a century earlier.

The Thames School charts discussed so far have been of relatively small scale and suitable for navigation over large areas of ocean or sea. But for western Europe there were also larger scale pilotage charts for the Channel and the east coast of England. In 1665 Comberford produced a matching pair of large four-panel charts of the Channel which were divided at the Straits of Dover: the one showing the waters west to Lands End and the tip of Brittany; the other for the area north to the North Sea between

---

[39] Howse and Sanderson, *The Sea Chart*, plate XVI; Waters, *The Art of Navigation*, plate L.

[40] Reproduced in Waters, *The Art of Navigation*, plate XLIX.

Fig. 8. Burston, Biscay-Channel, 1664. One of several somewhat different charts for the southwestern approach to the Channel. Soundings, similar to Hood's chart of 1592, are more numerous than usual on Thames charts. Greenwich, National Maritime Museum, G2 15: ¼ MS.

Fig. 9. Charts for the Northern Navigation matched those for the southwestern approaches. This one by Comberford, 1665 is one of three which he produced. A good clean example on the original hinged boards. Oxford, Bodleian MSDon.a4.

the Humber and the coast of Holland. The remaining two are by John Thornton of the English coast from the Thames to the Humber, 1667, and the Thames Estuary, 1682 (see Appendix 2). All four are on nearly the same scale of about two miles to the inch and show considerable bottom information, water depths, sand bars, shoals, and even some navigational aids. While pilotage charts were not new, particularly for the home waters near the English coasts, these are the earliest from the Thames School of which we have a record. But this type was to become increasingly important in the years after 1680 when the Thames School turned its attention increasingly to the Oriental Navigation.

The early Thames School practitioners did not ignore the Oriental Navigation although its chief concern was in the West. We know of ten charts dealing with ocean areas east of the Cape of Good Hope. Five of these adhered closely to a standard representation for the Indian Ocean which was clearly delimited from the Cape in the southwest along the east coast of Africa and Arabia, the south coast of Asia, Sumatra, Java and the west coast of Australia. Thornton's Indian Ocean, 1682 (fig. 10), typical of this representation, is a large four-panel chart drafted at the "oceanic" scale of about 45 leagues to the inch which was also used for charts of the Atlantic. Coolie Verner regards Thornton as the leading marine cartographer of his time,[41] and this chart is an example of his best manuscript style, carefully drafted, brightly colored, and well preserved. It is the last chart by Thornton of this type that we know about and may be said to be "climax Thames" in design, workmanship, and elaborate decoration. Half a century earlier Daniell had made two quite similar charts (1630 and 1637) that were much plainer and did not include the west coast of Australia despite the several Dutch landings there between 1616 and 1622.[42] Burston and Comberford each contributed one in the intervening period.

Four of the five remaining charts of the eastern seas are by Daniell—a pair for the Pacific Ocean, east and west, and two very similar ones for the China Sea and the east coast of Asia with a third by Comberford (Appendix 2). Daniell is thus the only early Thamesman whose surviving charts provide a balanced world coverage. The distribution of each chartmaker's signed output is generalized in Appendix 3 in terms of geographic dis-

[41] *The English Pilot, The Fourth Book* (TOT Facsimile; London, 1967), "Biographical Note," pp. v–ix, especially vii and viii.
[42] R. A. Skelton, *Explorers' Maps* . . . (New York, 1958), p. 211.

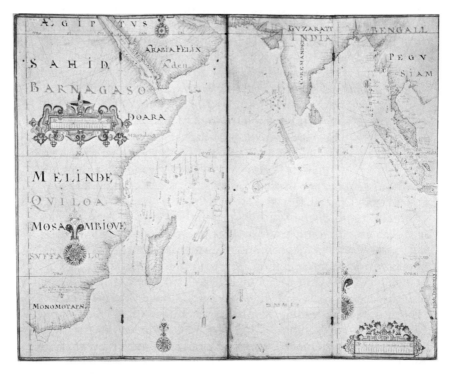

Fig. 10. Thornton, Indian Ocean, 1682, "climax Thames." One of the last oceanic navigation charts by the Thames School and a fine example of the more decorative style on four panels with hinges intact and matching scales on each half for convenience of use. By the 1680s Thornton was also engaged in the printed map and atlas trade. Oxford, Bodleian MS add. E.8.

tribution among seven major world areas. Daniell is represented in five of these with his surviving charts about equally divided—six charts on the Oriental Navigation and seven dealing with the West—Europe, the Mediterranean and the Atlantic. This geographic spread is also represented by the eight charts by Daniell now preserved at the Biblioteca Nazionale in Florence. This is one of the largest accumulations of the work of a Thames School chartmaker in a single repository and is believed to have been part of the working collection of Robert Dudley. Their dates, 1637 or 1639, even suggest that they may have been specially commissioned by Dudley for use in connection with the preparation of his own sea charts published in *Arcano del Mare* (see note 28).

The decade of the 1670s divides the history of the Thames School into two periods; several generalizations in regard to our survey up to this point can be derived from Appendix 3. Burston and Comberford both produced charts up to the year of their deaths, 1665 and 1670 respectively. Their passing removed the last two of the three chartmakers who had dominated the school for nearly sixty years. The first point to note is that, during the entire period from 1592 to 1670, there is a striking geographic concentration. Of the sixty-one charts presently identified from this period and noted on Appendix 3, forty-seven were of the Mediterranean, the waters of western Europe or the Atlantic. It follows that individual chartmakers also specialized. Burston is the prime example with six charts of the Mediterranean and three Biscay-Channel out of the ten that are attributed to him. Comberford's work was somewhat more broad, but still only two of his twenty-seven charts dealt with the Oriental Navigation. Only Daniell, as we have just noted, achieved a reasonable world coverage.

Also, it should be noted, there were definite periods of activity at least as represented by the surviving charts. First are the early "non-Draper" chartmakers (of whom Hood and Tatton are the most important) whose ten charts fall between 1592 and 1612. Then followed a hiatus of nearly a quarter of a century from which we have only three charts by Daniell, 1614, 1626, and 1630 and one by Comberford, 1626. A second period of activity seems to have begun in 1637, and we have listed forty-six charts by Daniell, Burston, or Comberford during the next quarter of a century. This represents less than one chart per year for each man, counting only the period of his active life. This is a low rate of production, especially since we know that all three were master chartmakers with four to six apprentices of record.

Two factors may help to explain this seemingly low productivity. First, sea charts have always been "used up" and subject to a high rate of loss. Those that have been saved are but a small part of the original corpus. But we have no way of establishing a relationship or comparison between the original product and the surviving remnant. Secondly, the chartmakers may not always have worked at that trade, and there is evidence that some of those on the master-apprentice tree were apprentices or even masters in other crafts. For example, Daniell took his freedom in 1590 but we have only three charts signed by him during the next forty-seven years, and the remaining ten charts were dated between 1637 and 1642 with his death following in 1649. Similarly, Comberford took his freedom in 1620 after an eight year apprenticeship with Daniell. His earliest surviving chart is dated 1626, but there is then a hiatus until 1641 at the beginning of his more active period. Meanwhile several apprentices had worked for him including John Burston, two of whose charts are dated in the 1630s. It seems probable that he was active during this period for which charts by his hand remain to be discovered. These long periods for which we have no surviving charts characterize the careers of other chartmakers. Sometimes there is an explanation, sometimes not.[43]

After 1670 there were changes as the Thames School was continued by the apprentices of Comberford, Burston, and Welch, and their apprentices, especially William Hack and John Thornton, as well as those who followed Thornton in the master-apprentice sequence. More chartmakers produced a much larger number of charts of different areas and type and with modifications of the traditional Thames School style. Ten chartmakers of record are shown on Appendix 3 to have produced over four hundred charts (another ninety are anonymous) in contrast to the sixty-one remaining from the years 1592–1670. In addition, of the ten there is only one chart that predates 1670: John Thornton's East Coast of England, 1667, a pilotage chart on two panels. This emphasizes the division in the development of the school.

With this later increase in productivity went a marked shift in

---

[43] Campbell pays attention to this problem but only for the Drapers and particularly in regard to Walsh and Daniell, "The Drapers' Company Chartmakers," pp. 88–93.

the area of interest. The Mediterranean was forgotten except for the single chart by John Thornton in 1679. Instead, major attention shifted to the Oriental Navigation and several hundred charts of the coasts of the Indian Ocean, East Indies, and eastern Asia were produced. Likewise Europe and the Atlantic received much less attention, but many charts were drawn for the New World, especially the coasts of North America and the Caribbean.

There was also a change in the type of chart. The traditional oceanic charts, on vellum and mounted on boards, were gradually superseded by larger scale pilotage charts for smaller coastal areas, some on vellum but none on boards. Paper was also used, especially by Hack, who bound many of his manuscript charts into atlases which provided systematic coverage of major coastal areas.

But the transition was gradual. "Old-style" navigation charts continued to be made. There are fourteen dating from the 1670s and 1680s, and we have already mentioned those by Welch, Gascoyne, and John Thornton. William Hack also worked in this form, and his earliest dated chart is a large four-panel chart of South America and the Caribbean, 1682 (fig. 11). Although similar to a portolan, it is more a map than a chart and very like the index maps that were included in the atlases of "The Great South Seas of America" of which Hack made a number of copies. He also produced the last board-mounted portolans we know of —a set of three dated 1687. They are large charts, each about 4' x 2'10", once mounted on two panels requiring three hinges along the long side. One is a world chart, which is the only one known by a Thames School cartographer. The second is a decorative and fanciful chart of the Galapagos Islands which is elaborately lettered, and dedicated to King Charles II, King James II, and various noblemen. The third is of the Caribbean.[44]

But the pilotage charts eventually displaced the older type as the main output of the Thames School. Several hundred survive from the period 1670 to the early eighteenth century when they, in turn, were displaced by printed charts and atlases. An output of such great magnitude cannot be reviewed in any detail in this

[44] The similarity in date, size, mounting, and general style of the three charts by Hack suggest that they may have been made for a particular purpose and that there could be more. All are at the British Museum:
   ... *Navigable Parts of the World* ... , *1687*, Add. MS 5414.6
   ... *Islands of the Gallipagos* ... , *1687*, Add. MS 5414.27
   ... *Jamaica Island & the Caribes* ... , (*1687?*), Add. MS 5414.25

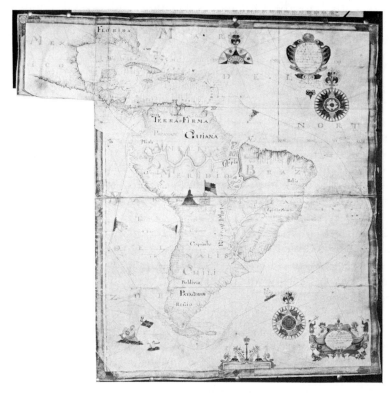

Fig. 11. Hack's chart of South America, 1682 was once on four panels and is traditional Thames style. It shows Sharpe's track of 1680–1681 and carries a dedication by Sharpe to the Duke of Albemarle as well as Hack's imprint. British Museum, Add. MS 5414.26.

short essay. Accordingly, I will refer mainly to John Thornton and William Hack, leading chartmakers of the period, who worked in somewhat different style and for different objectives.

John Thornton served his apprenticeship with John Burston and entered into the Drapers' Company in 1664. Mention has already been made of the four old-style charts that he made between 1667 and his classic "Thames Climax" chart of the Indian Ocean, 1682 (fig. 10). There followed another period of seeming inactivity, such as we have noted among earlier chartmakers. Only four manuscript charts of his making, including a large one of Hudson's Bay, are known between 1683 and 1698.[45] Both Thornton and John Seller, another member of the Thames School, made the transition from manuscript to printed charts and maps. During much of this period Thornton was associated with Seller and others in preparing charts and publishing the *English Pilot*. He was apparently connected in some fashion with the British East India Company for, in 1703, he published the *English Pilot Third Book* dedicated to the company and with 35 charts prepared for the Oriental Navigation.[46]

Thornton's largest output of manuscript charts appears to have been during the years 1699 and 1701. Twenty-two charts, over half his known output, are dated in either one of those years. They are on single sheets of vellum, show no signs of having been mounted, and all are for the Oriental Navigation. Three charts deal with India and the Persian Gulf. The rest are for the Far East: Malacca, East Indies, and the seas and coasts of East Asia. A good example is the Gulf of Siam, 1699 (fig. 12), which has a single compass rose and very subdued decorations and color by comparison with the one of the Indian Ocean, 1682. The single scale found on a number of these charts is also quite plain, without cartouche but with distinctive rounded box. Below it, also unadorned, is the simple imprint, "Made by John Thornton at the / Signe of the Platt in ye Minories / Anno Domy / 1699." Thornton included numerous soundings along the coasts, some from two much earlier voyages along both coasts of the gulf (see

[45] Three of the four are for the Oriental Navigation, all are in the British Museum:
Hudson's Bay, 1685, Add. MS 5414.20
Pegu, 1688, Add. MS 39178E
Achean (Sumatra), 1688, Add. MS 39178
West Coast of India, 1696, Eg. 741
[46] See Coolie Verner, "Engraved title-plates for the folio atlases of John Seller," in Wallis and Tyacke, eds., *My Head is a Map*; and Verner and Skelton, "Biographical Note" for *The English Pilot, The Third Book* (TOT Facsimile; London, 1970), pp. v–xiii.

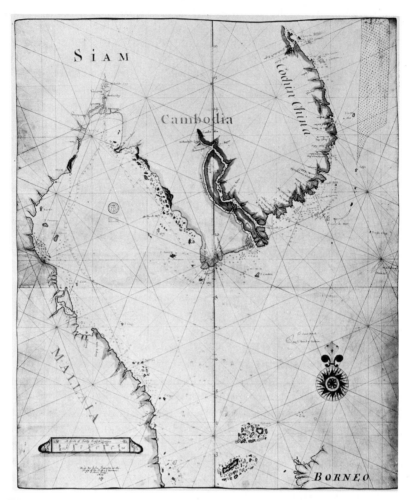

Fig. 12. One of John Thornton's pilotage charts for the Oriental Navigation. The distinctive scale with simple imprint beneath (lower left) is characteristic of several of these 1699 to 1701 charts by Thornton. The restrained style is in contrast to his chart of the Indian Ocean and the midcentury charts of Comberford et al. This is one of the seventeen presumed to have been taken by the French from an English ship in 1703. Paris, Bibliothèque Nationale, SH 181/0/6.

below). Augustine Fitzhugh, who was apprenticed to Thornton in the late 1670s, produced a nearly identical chart dated 1697. And both of these English charts are based on the work of Joan Blaeu, as can be seen by comparing his chart of 1664 (fig. 13) with the Thornton. Blaeu shows the same soundings in the gulf and attributes them to voyages of 1643 and 1644. Thornton and Fitzhugh both delete this note perhaps in an effort to "update" or to obscure their sources!

Seventeen of these charts are at the Bibliothèque Nationale in Paris, where Mlle La Roncière has studied them and commented on the similarity between several of them and charts in Thornton's *English Pilot, Third Book*, for which they presumably served as prototypes. But they also appear to have been used at sea, for she cites strong circumstantial evidence that these particular charts were taken by the French from an English East Indiaman, the *Canterbury*, off Malacca in December 1703.[47] Verner has also remarked on the similarity between some of these manuscript charts and those in the printed *Third Book*,[48] and both he and Mlle La Roncière have pointed to the close dependence of the English upon the Dutch charts, which is illustrated by comparing figures 12 and 13. There came to be standard representations for the Persian Gulf, Java Sea, Malacca, Sunda Straits, etc., and Thornton's charts of Tonkin Gulf are close copies of charts by Blaeu, complete with translations of the substantial notes on coastal and harbor conditions.

Another series of English manuscript charts for the Oriental Navigation, comparable to Thornton's and contemporaneous with his, are the fifteen charts signed by John Friend with dates from 1701 and 1709. These were discovered in 1969 at Chatsworth House in Derbyshire,[49] are on vellum, and show coasts and ports from the Persian Gulf to Amoy, about equally divided between India-Arabia and the Far East. John Friend completed his apprenticeship with Gascoyne in 1689. Friend's son Robert was apprenticed to his father in 1711, and his large chart of the South China Sea, 1719[50] is the latest work of the Thames School included in this inventory.

[47] M. de La Roncière, "Manuscript Maps by John Thornton, Hydrographer of the East India Company (1699–1701)," *Imago Mundi* 19 (1965), 46–50. Figure 1 is a reproduction of Thornton's chart of Sunda Strait, 1701.

[48] Verner, "Engraved title-plates."

[49] H. M. Wallis and W. P. Cumming, "Charts by John Friend Preserved at Chatsworth House, Derbyshire, England," *Imago Mundi* 21 (1971), 81.

[50] Newberry Library, Ayer Collection. The chart extends from Java and Malacca to the Philippines and Formosa with all the adjacent coast of East Asia. It is on two sheets of vellum with no signs of mounting.

Fig. 13. Joan Blaeu, Gulf of Siam, 1664. Prototype for Thornton's
chart of 1699 (fig. 12). Note the close coincidence of coastal details
and soundings, especially the tracks along each side of the Gulf.
Blaeu attributes them to a voyage of 1643, but Thornton omits the
note. Paris, Bibliothèque Nationale, SH Archives no. 17.

From this brief sequence it is clear that the engraved charts in the first atlases for the Oriental Navigation were closely related to the manuscript charts existing in the early eighteenth century. If the manuscript chartmakers copied each other, there is no reason to expect the engraver to do differently, especially when he also worked in manuscript as did Seller, Thornton, and the van Keulens. Consequently, the engraved charts at first, apparently, were no improvement over earlier manuscript examples of the same areas which they already possessed. This may be one reason why shipmasters and pilots continued to use the manuscript charts. Another reason, of course, is that it was not until 1703 that Thornton's *Third Book* provided systematic coverage, while the first Dutch printed sea-atlas for the Oriental Navigation was Part VI of van Keulen's *Zee-Fakkel*, first published in 1753.[51]

William Hack, in contrast to Seller and Thornton, did not produce printed charts and atlases for the trade. Instead, he concentrated on manuscript atlases for presentation to royal patrons and other important people. In sheer quantity he produced the largest number of charts of any member of the Thames School. Well over 300 are attributed to him and this represents a small part of his total output of manuscript charts in the years 1682 to 1702, a mass of material that has yet to be seriously studied. He is chiefly known for the several copies of the "Buccaneer's Atlas" (or *Waggoner of the South Sea*) which he copied from a Spanish *derrotero*, captured off the west coast of South America by Bartholemew Sharpe in 1680. Returning to England, Sharpe asked Hack to make a manuscript copy. This was done in 1682 and presented to King Charles II, who was much interested in the detailed information on the ports and coastal features of the Pacific coast of Spanish America from Acapulco to Cape Horn. Thus encouraged, Hack produced at least fourteen copies of his South Sea Atlas under various titles during the next twenty years. The individual issues contained charts ranging in number from 16 to 184 to a total of over 1,600[52]—enough

51 C. Koeman, *The Sea on Paper: The Story of the van Keulens and their Sea Torch* (Amsterdam, 1972), especially pp. 50–58. Koeman also reports that Gerard van Keulen produced 500 manuscript charts in the years 1706–1726 (pp. 21–22).

52 The most complete listing of Hack's atlases is by Thomas Adams, "William Hack's Manuscript Atlases of 'The Great South Sea of America,'" in the John Carter Brown Library, *Annual Report, 1965–66* (Providence, 1967), pp. 45–52. Also reprinted. Edward Lynam, *The Map Maker's Art* (London, 1953) contains

to have occupied a considerable shop, but there is no record of an apprentice serving under him. The charts, drafted on paper in bold colorful style, show profiles of coasts and harbors copied from the Spanish chart book and are not included in the Thames School materials.

But Hack's credentials as a functioning member of the Thames School are attested to by both training and output. The Drapers' records show that he began an apprenticeship with Andrew Welch in 1670, the year after Welch had completed a twenty-year apprenticeship with Comberford. There is no record that Hack entered into the freedom of the Company, and we have no chart signed by him and dated prior to 1682. But during these twelve years he must have done something to attract the attention of Captain Sharpe. His address, "At the Signe of Great Britaine and Ireland near new stairs in Wapping," placed him near the docks and other chartmakers. Here he produced the multiple copies of the *Atlas of the Great South Sea of America* and also a number of separate charts, mostly on vellum. Some of these were on wooden panels like the three large charts executed in 1687, which have already been mentioned (note 44). But mostly these were on single sheets of vellum, unmounted, and showed ports and coasts of North America and Caribbean Islands: Jamaica, 1682, Port Royal, 1683, North America north of Virginia, 1684, Darien, 1686, to name a few. His chart of Carolina, 1684 (fig. 14) is an example and shows his vigorous style. It is a plane chart characteristic of the Thames School, but Hack uses a square grid with a quarter rose instead of the rhumb network with decorated intersections so common on the standard portolan style; this square grid is characteristic of much of Hack's work and was also used by other members of the school.

Hack was the first to produce a number of charts of America in the Thames style,[53] but Fitzhugh, who was apprenticed to John Thornton, made several and there are a number of anon-

---

a chapter "William Hack and the South Sea Buccaneers," pp. 101–116. For a comparison of figure 11 with the index chart, see Campbell, "The Drapers' Company Chart Makers," plate 15, which reproduces one from a Hack atlas in the British Museum.

[53] The only detailed charts of parts of America by the earlier Thamesmen were:

Tatton, Guiana, 1608, BM Add. MS 34,320N

Comberford, Hispaniola, 1653, NMM G245.8/2MS

Comberford, South Part of Virginia, 1657, NMM G246.2/6MS. Another version New York Public Library, Map Room.

Fig. 14. Hack, Carolina, 1684. An example of Hack's separate maps of the coasts of North America, on vellum, unmounted. The square grid with only a quarter rose, vigorous style, and figured cartouches are characteristic of Hack's separate charts. British Museum, Add. MS 5415.G.5.

ymous ones yet to be identified. Hack's interest ultimately led him to produce an atlas of thirty-nine charts of North America and the Caribbean. The title page consists of a long, descriptive title in a heart-shaped cartouche: "A Description of the Coast, Islands, Etc. in the North Sea of America . . . by William Hack." There is neither date nor dedication and none of the charts is signed. But on chart ten of Ireland (a small island in the Bermudas) there is a descriptive note that refers to a survey of the island by one John Row in October, 1693. This provides a *terminus post quem* for the atlas and allows an approximate date within the last five years of the century. It is yet another of the Hack items contained in the King's Maritime Collection at the British Museum.[54]

Hack's most significant production in the Thames style was an atlas for the Oriental Navigation depicting coasts and ports of the Indian Ocean, the East Indies, and the China Sea. Three copies are known, two in the British Museum and one in the Library of Congress. As with the atlas of North America, the title pages often consist of a long descriptive title in the heart-shaped cartouche with Hack's name at the bottom. However, one is dedicated to Lord Somers, and it carries the date 1700.[55]

Each of the three atlases has between ninety and one hundred charts drafted on paper which are very similar, sometimes almost identical. The order of the charts differs somewhat, but the general arrangement is comparable between the three, and each one provides systematic coverage of the coasts of East Africa, Arabia, southern and eastern Asia from Cape of Good Hope to Amoy, and southern Japan. The scale is larger than Thornton's *English Pilot*, and these atlases of the Indian Ocean and China Sea may represent the first systematic coverage of such a large area of coast. Hack's chart of the Gulf of Siam (fig. 15) is clearly derived from the Dutch prototypes, although not as close a copy as are Thornton's. It is more of a chart than that of Carolina (fig. 14), 1684 with less detail on land and more underwater and coastal data for the pilot.

Finally, the substantial number of anonymous charts (ninety are indicated on Appendix 3) deserve comment. Most of them are at the British Museum or the Bibliothèque Nationale where

[54] British Museum, K. Mar VII.3,7 Tab 127, n.d. (1695–1700?).
[55] British Museum, K. Mar VI.1,1700, Dedicated to Lord Somers.
　　British Museum, K. Mar VI.1, n.d.
　　Library of Congress, Map Division, Phillips 3162, n.d.

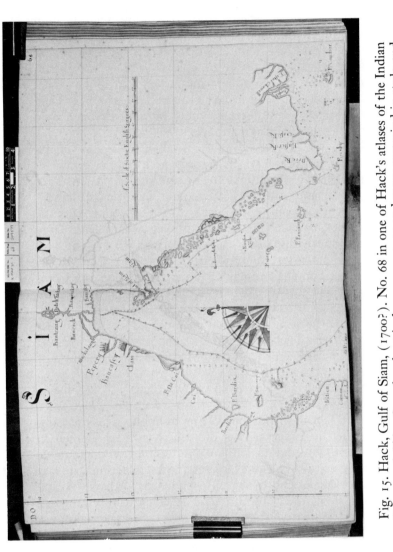

Fig. 15. Hack, Gulf of Siam, (1700?). No. 68 in one of Hack's atlases of the Indian Ocean and China Sea, the chart is drawn on paper and more restraind in style and decoration than his separate charts. Coast and soundings are shown in detail similar to Thornton and Blaeu. One of at least three sets of over ninety charts for the coastal areas from the Cape of Good Hope to southern Japan. British Museum, K. Mar. VI, 1.68.

they have been at least casually examined. They are pilotage charts for the Oriental Navigation or the New World and are quite similar to some that are signed by the Thames School chart-makers, post-1670. More detailed study may allow identification of the cartographers on the basis of hand or style. A number are unfinished and there are doubtless many others still to be examined. For example, a number of charts listed in the British Museum catalogs from their descriptions appear to be Hack charts from his atlas for the Indian Ocean. Since the individual charts are unsigned, identification would be difficult once they were separated from the atlas title page.

In this survey we have reviewed the work of the leading chart-makers of the Thames School. Its vigor is attested to not only by more than a century of activity by a sequence of chartmakers, but also by the quantity of the surviving output. The last decades of the seventeenth century constitute a particularly active period. The number of charts dated post-1670 is over eight times as large as the number surviving from the earlier period. Most of the later charts required much less time and effort to produce than did the larger and more complicated navigation charts of Burston, Comberford, et al., and this would reduce the disparity suggested by the 8:1 ratio. But, nevertheless, the four decades after 1670 still appear to be the most active period as indicated by the number of surviving charts. This brings up again the nagging question of the relationship between the chance of survival and the rate of production.

The incidence of survival is strongly influenced by chance entry into archival safety. Several groups of charts which have been discussed illustrate this point. For example, it has already been seen that the concentration of eight charts by Daniell in Florence is probably due to Dudley's acquisition of them about 1640 and their use as a chartmaker's reference collection. Similarly, the capture of the *Canterbury* in 1703 led, as we have seen, to the arrival in Paris of seventeen charts by Thornton within five years of their production, so that over half of Thornton's known manuscript output is now at the Bibliothèque nationale. This is a good deal less than the concentration at Chatsworth of fifteen out of sixteen known charts by John Friend, an "entrance into archival safety" that remains unexplained. The fourth example is William Hack who, unlike the others, himself provided for a high rate of survival by binding most of his charts into atlases that were presented to his royal and noble patrons so that they immediately enjoyed archival safety, such as it was. Another "accident of survival" is provided by the *Blathwayt Atlas* (see note 18), an assemblage of forty-eight maps and charts brought together about 1683 by the secretary of the Lords for Trade and Plantations. Among the thirteen manuscrapt maps in the *Atlas* is the only known work of James Lancaster, an apprentice of John Burston—a chart of Albemarle Sound dated 1679. In her study of the *Blathwayt Atlas*, Miss Black has identified and reproduced in facsimile nine additional unsigned charts and maps by Thames School copyists which, for the most part, reveal the use of recent information. They are also examples, some probably dating from the 1660s, of the growing emphasis upon larger scale pilotage charts of coastal waters and harbors which was to characterize the work of the Thames School in the late seventeenth and early eighteenth centuries.

A few comments can be made. First, it is clear that were it not for the circumstances leading to the early gathering of these several "collections" we would have many fewer charts from the Thames School. Secondly, they give a different indication of the level of productivity that was achieved by Thornton in 1699 and 1701 and by Hack over a period of about two decades. Such levels of productivity may well have characterized the work of the other Thamesmen whose output was dispersed and largely lost. This much is clear, what remains constitutes a substantial body of material which is only a portion, and probably a small one, of the original corpus. This disparity between original output and the surviving portion must always qualify our generalizations concerning the "output" of the Thames School.

Now let us turn from the charts to the chartmakers themselves and briefly consider three questions: how did they work, what was their socioeconomic situation, and what use was made of their charts? The only reason we can deal with these questions in a brief compass is that, so far at least, there is relatively little information that bears upon them.

The most complete set of instructions on chart making which was available to our seventeenth-century London practitioners was in Richard Eden's *Art of Navigation*, a sixteenth century translation of a manual on seamanship by Manuel Cortés, the famous Spanish navigator.[56] In this work, the second chapter of

---

[56] *The Arte of Navigation contenynyng a compendius description of the Sphere . . . Wrytten in the Spanythe tongue by Martin Curtes . . . Translated by Richard Eden* (London, 1561). Commander Waters in his *The Art of Navigation* comments at some length on Cortés; in this connection see especially pp. 62–63 and 75–77.

the third part consists of a twelve page discourse on "composition of Cardes for the Sea." The instructions for making the charts (cardes) must have been based on sixteenth-century practice as developed by the Mediterranean chartmakers. Examination of the Thames School charts indicates that they were made in this old-fashioned way. For example, Cortés specified definite colors for the rhumbs, with black for the eight winds (cardinal points of the compass), green or azure for the half, and red for the quarter winds—the Thames School charts conform. Upon the central intersection, or "chiefe compass" as he called it, in the center of the chart, Cortés directed, "gyue a prime . . . circle whiche may occupie in maner the hole Carde. This circle, some make with leade that it may be easely put out." And where this circle intersects the rhumbs radiating from the center, the chartmaker "shall leaue or make there other 16 compasses, euery one with his 32 wyndes . . . (and in some cases) paynt vpon the center of these compasses a flowre or a rose, with diuers colours and golde. . . ." In this manner Cortés described the standard arrangement of radiating rhumbs, secondary intersections, some with decorative roses exactly as they appear on the earlier charts of the Thames School and with modifications on the later ones by Thornton, Friend, et al.

The rhumb network, wrote Cortés, must be drawn first as a framework for the chart itself; on the Thames charts other line work, lettering, and even decorations "overprint" the rhumbs. Concerning the geographic detail of the chart, Cortés explained the use of thin tracing paper with soot-blacked "carbon" paper to trace and transfer coastlines, islands, and other features from an existing chart to a fresh sheet of vellum or paper. For cases where it was necessary to increase or decrease the scale, he illustrated the square grid method of transfer. Either method produced guidelines for coasts and other geographic details which were then inked-in by the draftsman. Careful examination of the Thames charts reveal places where a careless hand has left a bit of guideline uncovered. Instructions were also given for location of the coastal names, with red for the more important, and black ink for the lesser features; these conventions were followed by the Thames cartographers. Cortés's manual for the chartmaker is not extensive and the Thames craftsmen left no records but their charts. But whatever evidence on technique can be gleaned from an examination of the charts themselves is consistent with the instructions and sequence as set forth in Cortés.

The Thames School practitioners followed the traditional methods in the construction of the navigational charts, with some modifications when they turned to the larger scale pilotage charts in the latter part of the period.

When we turn to the societal situation of the Thames chartmakers or even their professional reputation, we also face a dearth of positive information. The charts themselves and the records at Drapers' Hall have provided what we know of names, dates, addresses, and the outline of master-apprentice relationships. But these sources, valuable though they are, are not elaborative, much less analytical. The chartmakers appear not to have left any accumulation of records or discourses on their work. Nor do they or their trade appear to have been of sufficient interest to merit much comment by contemporary observers and writers.

However, there is one piece of contemporary information: a record of a visit to Comberford's shop in 1655 which provides some insights into the status of the chartmaker and his position in the social and economic milieu of East London. The reasons for the visit are complex and noncartographic but are related to the prospect of an inheritance. Comberford was thought to be a rich old man with family connections and prospective heirs in Ireland. He turned out not to be rich, and William Dobbyns, who made the visit, entitled his account, "A Narrative of a Deceiving of Expectations." In it he recounts how he hired a coach to take him to the Tower and there found an old sea captain to guide him further. They took another coach which broke down, "the ways beyond the houses at Wapping being very deep." Dobbyns left the portly captain to the varied pleasures of a convenient tavern. He continued on foot through the mud and as he made inquiries, he was told "of one of such a name, which was a cardmaker or map-maker (that is one that draws maps of the sea-coasts for seamen), at which news I . . . began to fear for the success of my journey." He finally found Comberford's house and was met at the door by a woman (apparently the wife of Comberford's son) whom he described as "an ill-favored dirty slut."

In order to get the conversation going when he met Comberford, Dobbyns asked to see some maps "whereupon Comberford brought him into a low room and went to fetch some maps to show him. And upon debate of the price what a map was worth, 25 shillings was the lowest, and he swore he could make but one in three weeks and then he must work hard to, and that he and his son had much ado to maintain themselves and their family. . . ." The son joined them and in the course of the conversation mentioned that he knew some influential sea captains for whom he made charts. Then the conversation turned to non-cartographic subjects. Dobbyns shared a poor meal with unkempt children and slovenly adults and departed, having failed in his mission to find a rich uncle for his friends in Ireland.[57]

From this account we perceive that chart making itself was time-consuming, laborious, and ill paid even if we grant some exaggeration to the storyteller. Most revealing is Dobbyns's immediate reaction, on hearing that Comberford was a chartmaker, that he therefore would not be a man of substance. It would appear that chart making was not regarded as a very prestigious or remunerative occupation. Perhaps it suffered by comparison with the greater training, scientific accomplishment, and reputation attached to such callings as instrument maker, mathematician, or ship's captain.

A much more notable contemporary source is Samuel Pepys, whose interest in the improvement of sea charts is one aspect of his concern and responsibility for naval matters. Being a most knowledgeable and inquisitive Londoner, he was familiar with the work of the Thames School as well as with the printed chart trade, and these interests and concerns are recorded in diary entries and other notes. In the *Diary* the entry for July 22, 1663, includes the following, "Thence to my bookseller's, and find my Waggoners done. The very binding cost me 14s., but they are well done, . . . and so by water to Ratcliffe, and there went to speak with Cumberford, the platt-maker, and there saw his manner of working, which is very fine and laborious." This is tantalizing in its brevity and lack of detail from so keen and knowledgeable an observer. Pepys really tells us less than did Dobbyns eight years earlier, and he is more concerned about the binding on the volumes of his printed sea charts than impressed by Comberford's work. His reason for the visit is never stated, nor is Comberford mentioned again in the *Diary*. Two years later Pepys made arrangements with John Burston to draw three copies of "Lord Sandwich's chart of Portsmouth Harbor," one for the king, another for the duke of York and another for himself. In February and March of 1665 Pepys reported several trips to Burston's shop on Ratcliff Highway, in this connection, and

[57] Great Britain Historical Manuscripts Commission, *Egmont Papers*, vol. 1, part 2 (London, 1905), pp. 571–573.

that the charts were actually completed.[58] Unfortunately, none of these charts has survived. There is none in the Pepys Library, and Robert Latham, the librarian, reported lack of success from his inquiries at likely repositories, including the private collection of the earls of Sandwich.[59]

Pepys also had casual contact with Gascoyne and Thornton. His library contains two charts by Thornton. One is a copy of an Elizabethan port sketch, the other a simple and non-Thames style chart of European coasts. Pepys seems not to have acquired any Thames School charts because none is listed in his "Catalog of my Books of Geography and Hydrography"[60] compiled a few years before his death. By contrast the catalog did list several printed atlases of sea charts, Dutch, French, and English. So, even though Pepys was well aware of the Thames plattmakers and had contact with several members of the school over an extended period, he was little concerned with their charts and availed himself of their services only as copyists.

There can be no doubt that manuscript charts were used at sea. The mere fact of the continued production of hand-drawn charts into the eighteenth century is a general indication of demand and utility. This is true of the French, Italian, and especially Dutch chart trade, as well as the English. But hard evidence is difficult to accumulate.

An indication of the use at sea are the complaints, recorded in logs and journals, concerning the accuracy of the charts, usually without identifying the chart or its maker. After all, the pilot or captain was not writing his log for posterity and a simple notation of a particular error "on our platt" was sufficient identification for his purpose, especially if the notation was entered on the chart itself. Sometimes, however, cartographers were identified as in the case of Walter Payton's second voyage to the East Indies in 1614, where errors in the representation of the African coast on "Daniels Plats" were specified and compared with the greater accuracy of the "Plano of Tottens" (read Tatton?).[61]

[58] Wheatley, *The Diary of Samuel Pepys* (Cambridge, 1904–1905), III, 204; IV, 333, 340, 342, 343, 346, and 394.

[59] Robert Latham, personal correspondence, 3 June 1971.

[60] Pepys Library, #1296, #2940 and #2700.

[61] *Purchase His Pilgrims* (Glasgow, 1904), IV, 291. Assuming that Payton was referring to charts by John Daniell and Gabriel Tatton, the reference is interesting on two counts. The specific areas referred to were East Africa and Madagascar. There is no known chart by Tatton showing this area and the earliest by Daniell is his Indian Ocean 1630. So Payton must have been using charts now lost. For Daniell this would have been another chart dated 1614 or earlier, at the very beginning of his known career as a chartmaker.

Seventy years later, Pepys provided another example. On a voyage to Tangier, in 1683, he organized an exercise in navigation by having a number of navigators keep course plots, some on hand-drawn and some on printed charts. The former were still in use but the latter were more accurate according to Pepys, but Pepys does not identify either.[62]

Neither general considerations nor contemporary references to manuscript charts used at sea are as compelling evidence as the survival of charts that show evidence of having been used for navigational purposes. A number of the charts have been separated from their wooden panels or stained at least by dampness, if not water, and in a few cases show signs of wear on the scales as if used for frequent measurements of distance. But these evidences could be the result of use (or misuse) on land as well as on shipboard. It is the survival of charts with plots of ships' positions and tracks during actual voyages which provide the firmest evidence. There are several examples.

But even these may sometimes be misleading. On Hood's chart of the Western Atlantic 1592 there is a clear track of a vessel which can be seen, (fig. 3); it extends from the coast of South America, through the Antilles and recurves to the northeast out into the Atlantic. The chart was in Dudley's possession and the track is similar to that described in the account of his voyage to Guiana in 1594. But it seems likely that the plot was added later in his study, rather than on shipboard.[63]

Better examples are provided by two late seventeenth-century charts of the Indian Ocean by the prolific cartographer Joan Blaeu which show well-defined navigational tracks and indubitable evidence of use at sea. They are in the Bibliothèque nationale where I have had an opportunity to examine them and where Mlle Foncin has briefly described them and noted the tracks. The first chart, dated 1687, is a standard chart of the Indian Ocean unmounted and much plainer than the Thames charts of the same area and vintage. The tracks are most visible in an east-west alignment between the Straits of Sunda and the southern tip of the Maldive Islands. One is marked by dots with small freehand circles for emphasis. Another is in the form of compass arcs with a dot for position. The arcs are convex to the west,

<hr />

[62] Edwin Chappell, ed., *The Tangier Papers of Samuel Pepys* . . . , Publications of the Navy Records Society, 73 (London, 1934), 126–128.

[63] Alan Stimson, National Maritime Museum, Greenwich, personal correspondence (October 1971).

indicating a westbound course. To the north of the Maldive Islands traces of tracks can be picked out on the original chart along the coasts of India, Ceylon, Persia, and in the Bay of Bengal.

The second Dutch chart, also by Blaeu and dated 1698, is too faint and dark for effective reproduction. Two tracks are shown between the Cape of Good Hope and Sunda—the eastbound one (dot and circle representation) running southeast to the latitude of the strong westerlies, then east with the wind until it was necessary to run northeastward toward Sunda. The westbound course, marked by compass arc and dot, follows a more direct route, Sunda to the Cape, and to the north of the eastbound track. Another feature is that the chart appears incomplete at first glance. It does not show the entire coastline but only headlands, islands, and coastal segments important to Indian Ocean navigation and particularly the Sunda-Cape run. Mlle Foncin has noted this feature and suggested that it represents an effort of the Blaeu firm to simplify manuscript charts and mass produce them for active navigational use.[64]

The fourth "track chart" is a Thames School chart of the South China Sea which can be definitely attributed to John Thornton. It is at the Bodleian Library and, like several others in that collection, has been mutilated by cropping along both sides which has removed the author imprint and most of the scale. The end that remains is identical to the scales found on Thornton's manuscript maps of the Oriental Navigation, 1699 and 1701, and there are other similarities in style and design which support attribution to Thornton and a date of c. 1700.

The chart is dirty and worn, and the track, faintly drawn in pencil, is just discernible on the original. It begins off the mouth of the Mekong River and follows a curving course northward toward Macao. Some of the plotted positions are dated June 20 to 27, but the year is not indicated. That this chart was used at sea can hardly be doubted, most likely in the first years of the eighteenth century or about the time that the French took the seventeen Thornton charts from another English merchant ship.

Thornton's substantial output of manuscript charts in 1699 and 1701 had at least two uses. It provided prototypes for some of the printed versions in his *English Pilot, Third Book*, published in 1703. It is also clear that some were for use at sea. At the

[64] M. Foncin et al., *Catalogue des cartes nautiques*, pp. 142–143.

same time, there is evidence that the Dutch were actively producing manuscript charts on vellum for navigational use. In fact, the first Dutch printed atlas with charts for the Oriental Navigation did not appear until half a century later.[65] So it appears that manuscript charts remained competitive for longer in the Oriental Navigation than they did in the West. Of course, eventually they were displaced by printed atlases and charts, as had happened earlier in European and Atlantic waters.

The history of the Thames School of chartmakers can thus be traced rather fully except for the identification of the prototypes of the earliest Thames charts made in the 1590s. But a century later we find Thornton and Friend copying Dutch charts line by line and word by word. Pepys's frequent complaints that English printed charts were but copies of the Dutch applied also to the manuscript charts. The English chart trade of whatever type still had to free itself from dependence on the Dutch.

[65] C. Koeman, *The Sea on Paper*.

Appendix 1

PLATT-MAKERS OF THE DRAPERS' COMPANY—LONDON—1590–1719
A DIAGRAM OF MASTER-APPRENTICE RELATIONSHIPS AND DATE THEREOF AS FOUND IN THE RECORDS OF THE WORSHIPFUL COMPANY OF DRAPERS

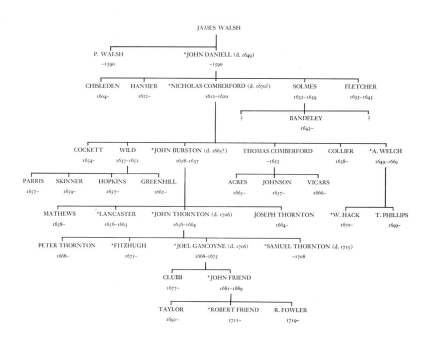

*Persons known to have produced at least one Portolan-Type Chart.
Date of Apprenticeship 1604—1665 Date of Freedom.

Appendix 2

THAMES SCHOOL OCEANIC CHARTS BY AREA

Abbreviations:
BM   = British Museum
NMM = National Maritime Museum

Mediterranean

Tatton, c. 1600, Newberry, Ayer MS Map 22.
Lupo, c. 1600, BM, Add. MS 10,041.
Reynolds, c. 1612, Florence, Bib. Naz., Port. 11.
Comberford, 1626, Private Collection.
Burston, 1638, NMM, G.230:1/15 MS.
Burston, 1640, BM, Add. MS 19,916.
Daniell, 1642, Mantova, Bibl. Governativa, MS A.U.6n136.
Daniell, 1642, Dublin, Trinity Library, MS 1209 #81.
Burston, 164?, BM, Map Room, C21 E2.
Burston, 1647, Milan, Crespi Library, reported 1896, now lost.
Comberford, 1647, Yale, Sterling Library, Map Room.
Comberford, 1657, Dartmouth, Baker Library.
Comberford, 1657, BM, Add. MS 5415.C.2.
Burston, 1658, NMM, G.230:1/5.
Burston, 1659, NMM, G.230:1/13.
Comberford, 1663, BM, Add. MS 26,665.
Comberford, 1664, BM, Add. MS 5414.8.
Comberford, 1666, University of Kansas, Spencer Library, Summerfield MS J 7:1.
John Thornton, 1679, BM, Add. MS 5414.7.

North Atlantic

Hood, 1592, (North America-Caribbean)* Munich, Staatsbibliothek, Cod. icon. 140 Nr. 84.
(Hood), (159?), BM, Add. MS 17,938B.
Tatton, 1602, Florence, Bib. Naz., Port. 21.
Daniell, 1639, Florence, Bib. Naz., Port. 13.
Comberford, 1646, Florence, Bib. Naz., Port. 25.
Comberford, 1650, NMM, G.213:2/2.
Comberford, 1657, BM, Add. MS 5414.3.
Comberford, 1659, (North America-Caribbean)* Harvard, Houghton Library, 51–308.
Anon., Early XVII, (Half-Chart, East), Huntington, HM 2098.
Welch, 1674, NMM, G.213:2/5.
(Welch), (167?), (Half-Chart, West), Huntington, HM 43.
(Welch), (167?), (Half-Chart, West), BM, Add. MS 31,858.
Gascoyne, 1678, (Half-Chart, West), NMM, G.213:4/2.
Gascoyne, 16(8)6?, BM, Add. MS 5414.3.

* The "North America-Caribbean" charts by Hood and Comberford are complete in themselves and not "half-charts" although they are very similar to the western half of the other North Atlantic charts on the list. Also, they show no evidence of having been mounted on boards, an unusual feature for Thames School charts.

West European Waters

Southwestern Approaches

Hood, Biscay-Channel, 1596, NMM, G.224:1/2.
Daniell, Azores-Channel, 1626, BM, Add. MS 18,664B.
Comberford, Biscay-Channel, 1641, University of Kansas, Spencer Library, Summerfield MS J 7:2.
Burston, Azores-Channel, 1660, Oxford, Bodleian, MS Add. E.10.
Burston, Biscay-Channel, 1664, NMM, G.215:1/4.
Burston, Biscay-Channel, 1665, NMM, G.215:1/2.
Comberford, Biscay-Channel, 1666, NMM, G.215:1/3.
Comberford, Biscay-Channel, 1668, University of Kansas, Spencer Library, Summerfield MS J 7:3.

Northern Navigation

Daniell, Northern Europe, 1637, Florence, Bib. Naz., Port. 14.
Comberford, Northern Europe, 1651, NMM, G.213:3/1.
Comberford, North Sea-Baltic, 1665, Helsingor, Danish Maritime Museum, 121:49.
Comberford, Northern Europe, 1665, Oxford, Bodleian, MS Don. a.4.

Pilotage Charts

Comberford, Channel, N.E., 1665, Yale, Sterling Library, Map Room.
Comberford, Channel, S.W., 1665, BM, Add. MS 41,650.
Thornton, English Coast, Humber to Dover, 1667, NMM, G.218:10/6.
Thornton, Thames Estuary, 1682, NMM, G.218:8/7.

After this essay had been set in type a fifth pilotage chart was brought to my attention by Dr. Gunther Schieder, Utrecht. It is comparable in scale, date, and style to the four listed here: Burston, Channel, N.E., 1666, Amsterdam, Maritime Museum, Z 11 a–1–88.

South Atlantic

Daniell, 1614, BM, Add. MS 5415.C.1.
Daniell, 1637, Florence, Bib. Naz., Port. 8.
Comberford, 1647, BM, Add. MS 31,320B.
Comberford, 1664, BM, Add. MS 5414.2.
Comberford, 1670, Oxford, Bodleian, MS Add. E.9.
Thornton, 1681, Library of Congress, Map Division, LC #18.

Indian Ocean

Daneill, 1630, BM, Add. MS 18,664A.
Daniell, 1637, Florence, Bib. Naz., Port. 10.
Burston, 1665, Paris, Bib. Nat., SH 213/3/4.
Comberford, n.d., BM, Add. MS 5414.11.
Thornton, 1682, Oxford, Bodleian, MS Add. E.8.

China Sea and Pacific

Daniell, China Sea, 1637, Florence, Bib. Naz., Port. 12.
(Daniell), n.d., China Sea, BM, Add. MS 5415.1.1.
Comberford, China Sea, 1665, NMM, G.256:1/1.
Daniell, East Pacific, 1639, Florence, Bib. Naz., Port. 23.
Daniell, West Pacific, 1636, Florence, Bib. Naz., Port. 24.

Appendix 3

THAMES SCHOOL CHARTMAKERS WITH OUTPUT BY PERIOD AND AREA

	Total #	Range of Dates	Mediterranean Sea	West Europe and Atlantic	America North	America Carib & South	Indian Ocean South Asia	Southeast & East Asia	Other
HOOD	4	1592–1604		4					
TATTON	4	1596–1608	1	1		1			Pacific
REYNOLDS	1	(1612)	1						
LUPO	1	(162?)	1						
DRAPERS									
DANIELL	13	1614–1642	2	5			2	2	2 Pacific
BURSTON	11	1638–1665	6	4			1		
COMBERFORD	27	1626–1670	7	15	2	1	1	1	
ANONYMOUS	1	?		1					
Subtotal to 1670	62		18	30	2	2	4	3	3
J. THORNTON	33	1667–1701	1	3	2		9	18	
WELCH	6	1674–1677		3			2	1	
GASCOYNE	4	1678–1686		2	1		1		
HACK	331	c. 1680–?			41	12	155	122	World
FITZHUGH	9	1683–1697			3		4	1	Guinea
J. FRIEND	16	(1703–1709)					8	8	
S. THORNTON	1	1707						1	
R. FRIEND	1	1719						1	
J. SELLER*	2	1618–?					2		
LANCASTER	1	1679			1				
ANONYMOUS	90	c. 1680–?		1	16	20	39	14	
Subtotal from 1670	494		1	9	64	32	220	166	2
TOTAL	556		19	39	66	34	224	169	5

* Seller was a member of the Merchant Taylors, not the Drapers' Company.

III

MAPPING THE ENGLISH COLONIES IN
NORTH AMERICA: THE BEGINNINGS

Jeannette D. Black

All who have worked with seventeenth-century English maps are acutely aware of what is missing. Time and again references are found in the contemporary records to maps that are not at present known to have survived. There are reports containing extensive descriptions of one area or another which must originally have been accompanied by maps, but they are nowhere to be found. Many of them have disappeared for good. However, there are maps that survive for which no documentary explanation has yet been found. It is certain that there are maps still surviving unrecognized in libraries and in private collections, still waiting to be identified. No one has ventured to estimate the rate of survival of these maps, but certainly the percentage is low.

A case in point is Nicholas Comberford. Not many years ago it was customary to speak of "the" Comberford map, meaning a manuscript in the New York Public Library showing northern Carolina and southern Virginia.[1] Now, we know of nearly thirty maps of Comberford's making, and even this number may be only a sampling of what Comberford must have produced in his fairly long professional career. The same is true of printed maps, although in somewhat lesser degree. For instance, in 1911 when P. Lee Phillips published a short monograph on Augustine Herrman's 1673 map of Virginia and Maryland, only one copy, in the British Museum, had been found.[2] In 1932 another copy was

[1] William P. Cumming, *The Southeast in Early Maps*, no. 50 (Chapel Hill, 1962).
[2] P. Lee Phillips, *The Rare Map of Virginia and Maryland* (Washington, 1911).

acquired by the John Carter Brown Library.³ Then in 1949 it
was discovered that there were three copies in Paris in the Bibliothèque Nationale. An exchange for one of these was arranged,
and it is now in the Library of Congress.⁴ More recently still
another copy has been found in a volume of maps that once belonged to Samuel Pepys and has been among his books in Magdalene College, Cambridge, for nearly three centuries.

These matters are mentioned simply to suggest why there are
dangers in making general statements about seventeenth-century
English maps and should be borne in mind when considering the
generalizations that follow.

The period of the late sixteenth and much of the seventeenth
century has been called, and rightly so, the golden age of Dutch
cartography. The cartographers and publishers of Amsterdam,
especially, were leaders in supplying maps for all of western
Europe, including England. Even for representations of their
own coasts and harbors, Englishmen depended on Dutch maps.
English cartographers lagged behind their practical navigators,
and the engraver's craft, through which maps were brought to
the attention of the public, spread even more slowly. Admiration
for the Dutch product was a habitual, almost a colonial attitude.
Even after the Anglo-Dutch Wars the attitude was slow to
change. Not without some natural irritation, Samuel Pepys wrote
in 1683, "The Dutch print our very bibles for us."⁵ The year
before, an English map publisher advertised in a newspaper,
"And observe that Rob. Walton scorns to counterfeit Dutch
Maps, as Jo. Overton hath, and put the Dutch names; only as I
conceive, to cause people to think they are so; whereas they are
but counterfeits and not so good, as the most ingenious know;
but I will sell the true ones at the same price he sells his counterfeits."⁶ In the course of his voyage to Tangier, Pepys recorded
conversations with Thomas Phillips, a military engineer who had
had considerable experience in chart making. Phillips showed him
a book of John Seller's (presumably the first part of *The English*

<hr>

³ Facsimile published by the John Carter Brown Library, Brown University
(Providence, 1941; reprinted 1948 and 1958).

⁴ Walter W. Ristow, "Augustine Herrman's Map of Virginia and Maryland,"
in *Library of Congress Quarterly Journal of Acquisitions*, August 1960, pp.
221–226, reprinted in *A la Carte* (Washington, 1972), pp. 98–101.

⁵ Edwin Chappell, ed., *The Tangier Papers of Samuel Pepys*, Publications of
the Navy Records Society, 73 (London, 1935), p. 106.

⁶ *The True Protestant Mercury: or Occurrences Foreign and Domestick*
(May 24–27, 1682).

Pilot, 1670), which had "the very same platts with the Dutch without a Dutch word so much as turned into English, much less anything in the maps altered. And he [Phillips] says that he knows it to be true and Seller will not deny it, that he bought the old worn Dutch copper plates for old copper, and had them refreshed in several places, and has used them in his pretended new book. He says that the Dutch have laid down their own coast so as is never to be bettered. . . ."[7] There was no reason for Seller to deny using old plates when he had no better maps to substitute for them—he was simply following the custom of many Dutch publishers of maps and atlases. The royal authorization for his publications gave no indication that their content was to be entirely new but stated they were to be "collected" and composed by him. His contribution with these earlier books was to be that they were published in England rather than imported from the Netherlands. Even this was a beginning, and by the 1670s the English were ready to promote their own publication of maps.

The seventeenth century marked the beginnings of English colonization in North America, and the maps necessary for most of this extensive activity could not be had from the Dutch. By 1600 the English had been in Newfoundland and the part of the Carolina coast which had been the scene of Raleigh's unsuccessful colonial ventures. But before the century was over, they not only had spread their settlements along the east coast between Canada and Florida, but they were in Hudson's Bay and on a number of West Indian Islands. They had even been briefly on the northern coast of South America. Of course during the years of the New Amsterdam colony, some Dutch maps of the east coast were produced, and they were used by Englishmen along with those they were beginning to make for themselves. But the Dutch maps of North America were not a large or distinguished group. In fact, if one considers the Dutch maps of America only, there never would have been a golden age for the Dutch mapmakers. Their chief field of new endeavor lay elsewhere, in the more immediately lucrative continents of the South and East. For North America the English were to a great extent on their own.

---

[7] Chappell, ed., *The Tangier Papers,* p. 107. Further information on Phillips is given in A. H. W. Robinson, *Marine Cartography in Britain* (Leicester, 1962), pp. 88–89.

At the beginning of the seventeenth century the best English cartography produced the picture of North America as shown on the so-called Wright-Molyneux map of the world in Richard Hakluyt's *Principal Navigations*, 1598–1600 (fig. 1). In it the West Indies are recognizable, taken from continental sources. Newfoundland and the St. Lawrence were known from the French, but there is no Hudson's Bay, and there is almost nothing in the interior except a rumor of lakes. The east coast, north of Florida, has almost no resemblance to reality. By the end of the seventeenth century London was producing compiled maps of what they were already calling the English Empire in America, showing considerable detail, and with a commendable, if still incomplete, approach towards accuracy. Most of the development during the century is traceable, in one way or another, to the English colonizing activities.

The beginning of English mapping in the new areas was the result of necessity. The English made maps because they had to, and conversely they seldom made maps for which they did not feel an immediate need. Nothing was done for pleasure, nor was there much in the way of pure geographical scholarship. It was a serious business. Even the map made in the 1670s by a fourteen-year-old boy named Cotton Mather, which was sent to his uncle in Dublin, was probably a serious production. But unfortunately it is one of the maps that has not survived.[8]

The colonizers needed first reconnaissance maps, studies of the coast that would show where to find the best and most likely places for planting their colonies. After the sites were chosen and settlement planned, they needed maps to show where the colonies were, locational maps that could show prospective settlers and investors where they might go, sometimes what they might find there, and who their neighbors would be. Later, when the need developed and was recognized, there were to be maps of the colonies themselves, made by settlers for their own use and for the use of their backers in England, and later for the use of the governmental agencies that slowly, as the century wore on, developed into tools of a more or less efficient colonial administration.

These three purposes do not constitute a classification of maps. They represent no chronology except in the cases of individual colonies, and of course many maps served more than one of these

---

[8] Nathaniel Mather, Dublin, February 26, 1676, to Increase Mather, Boston, *Collections of the Massachusetts Historical Society*, ser. 4, vol. 8 (1858), p. 9.

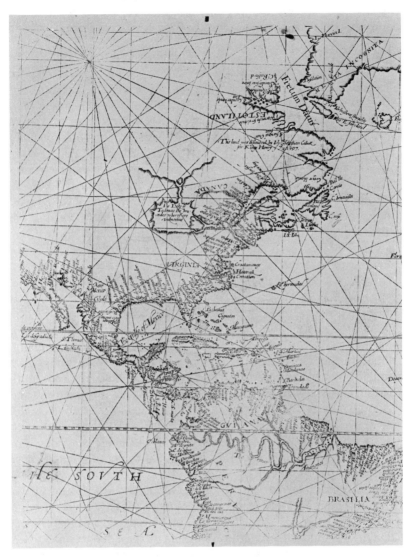

Fig. 1. Detail of world map in Richard Hakluyt, *The Principal Navigations*, London, 1598–1600. From copy in the John Carter Brown Library.

three purposes. The intention, however, is useful as a guide to understanding early maps, explaining why they look the way they do, why they show some things, and why they leave out others.

An early example of a map combining the purposes of reconnaissance and location was John Smith's well-known map of Virginia.[9] It showed the results of his initial explorations of Chesapeake Bay as well as the location of the settlement already begun at Jamestown. After Smith severed his connection with the Virginia Company, however, he made a map of New England which was strictly a map of reconnaissance (fig. 2). In 1614 he made a voyage to the New England coast with two ships sent out by some London merchants, not for the purpose of exploring or making maps, but with purely commercial objectives. First, they were to catch whales, second, they were to look for mines of gold or copper (a recurring theme with the trading voyages of the early period), and if these activities were not successful, they were to catch fish and trade for furs with the natives in order to recover the expense of fitting out the voyage. Smith's purpose of searching out a good place for a colony was entirely his own idea. In the months of June and July 1614, he ranged the coast from Penobscot Bay to Cape Cod. He was not the first to do this by any means, nor did he make such a claim. Among others, his friend Bartholomew Gosnold had followed a similar course a few years earlier, but he left no map, and Smith had little to help him in his map making. As he expressed it (and his own words are better than any paraphrase), "their true descriptions are concealed, or never well observed, or died with the Authors: so that the Coast is yet still but even as a Coast unknowne and undiscovered. I have had six or severall plots of those Northern parts, so unlike each to other, and most so differing from any true proportion, or resemblance of the Countrey, as they did mee no more good, then so much waste paper, though they cost me more."[10] By "plots" Smith may have meant English navigators' sketch maps, although none of these have come down to us, but he was also certainly familiar with the Wright-Molyneux map. For his purposes it was certainly waste paper,

[9] First published Oxford, 1612, and included later the same year in *A Map of Virginia. With a Description of the Countrey.* Reproduced as Library of Congress Facsimile no. 1, 1957.

[10] John Smith, *A Description of New England* (London, 1616), pp. 4–5. The New England voyage is discussed in chaps. 21–22 of Philip L. Barbour, *The Three Worlds of Captain John Smith* (Boston, 1964).

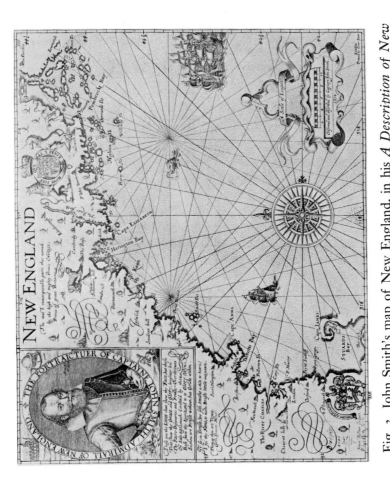

Fig. 2. John Smith's map of New England, in his *A Description of New England*, London, 1616. From copy in the John Carter Brown Library. Reduced.

although in our day no dealer would put it in that class. Smith says further,

I have drawen a Map from Point to Point, Ile to Ile, and Harbour to Harbour, with the Soundings, Sands, Rocks, & Land-marks as I passed close aboard the Shore in a little Boat; although there be many things to bee observed which the haste of other affaires did cause me omit: for, being sent more to get present commodities, then knowledge by discoveries for any future good, I had not power to search as I would: yet it will serve to direct any shall goe that waies, to safe Harbours and the Salvages habitations.

Smith's map was engraved and published two years after he made his coastal survey as the illustration for his *Description of New England*. We do not have his original manuscript, and there is good reason to believe that Smith did not have it either. During the two years between his voyage and the printing of the map, he had set out on two other voyages, having persuaded backers to go along to some extent with his idea of founding a colony. Neither of these expeditions, inadequately manned and equipped, ever reached New England. On the second voyage he was taken prisoner by a French privateer, made a dramatic escape alone to the coast of France, and finally returned to England to pursue the struggle to find someone to back his colonial dream. His books and papers were not with him when he was captured. The text of his book was written from recollection while he was a prisoner. Unless Smith had left his map in England after returning from the original voyage, and unless he found an opportunity to seek it out two years later when he came home from France—and this seems unlikely—the map as well as the book was done from memory. His original manuscript was probably considerably better and more detailed. It probably included soundings, which he explicitly says he noted. If his map was done from memory, however, it must be rated as a remarkable performance. It shows readily recognizable features along an intricate stretch of coastline for which he had no map to serve as his model. Soundings could hardly have remained in his memory for two years, and perhaps he should be given credit for not having tried to fake them.

The names on Smith's map are so curious that they have served to take attention away from his delineation of the shoreline. Only a few of the names became permanent, at least in the places where they are on the map. Smith was by necessity, as well as by natural talent, a public relations man. He put three dedications into his New England book, one of them to Prince Charles (later Charles I), then fifteen years old. Smith says he asked the prince to change the Indian names on the map to English names, and he claims Charles was responsible for a long list of them. Historians have questioned this, pointing out that there is no solid evidence that Smith ever had an audience with the prince, but their objecton has little meaning because it could have been done by letter or through some intermediary without leaving any permanent record. The psychological tendency to give familiar names to unfamiliar places was well known to Smith, and certainly the prince would have been likely to suggest names like Falmouth, Ipswich, Dartmouth, Cambridge, and Plymouth. Some of the names were later attached to actual settlements, mostly in quite different places.

Maps of reconnaissance were still essential much later in the seventeenth century. In 1662 the towns of Charlestown and Newbury, near Boston in Massachusetts, were already thoroughly settled, and space for expansion was limited. Some of the townspeople sent a small sailing vessel to Cape Fear on the coast of Carolina, where as yet there were no English settlements, to investigate the possibilities for relocating themselves. The report of the voyage remained in manuscript unrecognized until recently. It was a glowing account which inspired the sending out soon after of a group of colonists with their cattle and all the paraphernalia for beginning a new settlement. For various reasons, some of which are known and some not known, the settlement was abandoned before it was begun. The only version of the account of the voyage known to survive has no map. There are, however, two contemporary manuscript maps that illustrate it. One is an attractive colored copy, probably made in the 1670s (fig. 3); the other, also undated, is a pen-and-ink sketch in the hand of the philosopher John Locke, a protégé of the earl of Shaftesbury, who was one of the proprietors of Carolina (fig. 4). The two copies are superficially very different in appearance, but their content is essentially identical, and they were probably from the same original. Locke's version gives more information as to its origin. The heading reads: "Discovery made by William Hilton of Charles Towne In New England Marriner . . . layd down in the forme as you see by Nicholas Shapley of the town

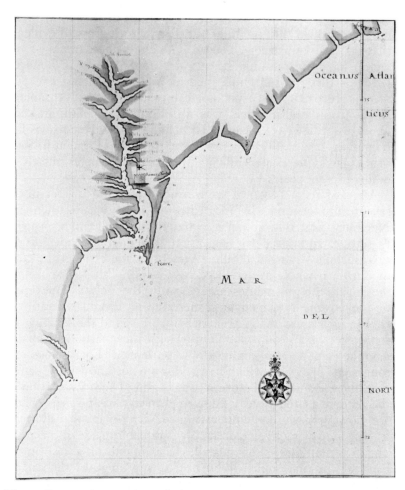

Fig. 3. Decorative version of chart of Cape Fear River by Nicholas
Shapley, 1662. Manuscript in the Blathwayt Atlas (Map 19), in the
John Carter Brown Library. Reduced.

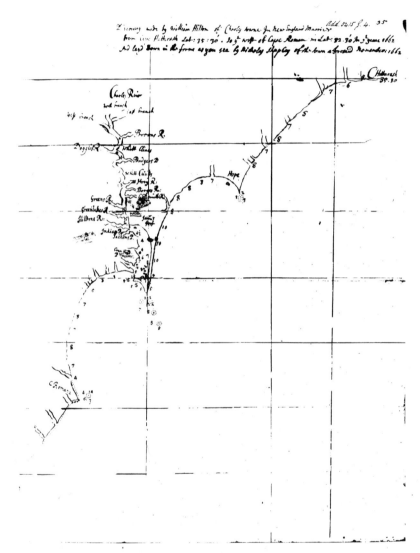

Fig. 4. John Locke's pen-and-ink version of chart of Cape Fear River by Nicholas Shapley, 1662. Manuscript in the British Museum (Add. MS 5415.g.4). From photograph in A. W. Hulbert, *The Crown Collection*, (Cleveland, 1904–1908), Vol. V, no. 30. Reduced.

aforesaid November: 1662." Shapley was apparently not aboard the vessel commanded by Hilton but was a retired mariner, "clarke of the writts" in Charlestown, just across the river from Boston. When he died a year or so later the inventory of his estate included various compasses and other instruments as well as nautical books and a stock of parchment. This leads us to suspect that he had a shop similar to those of the "mathematical practitioners" of London.[11] At any rate his services were called upon to put into acceptable form the information from Hilton's log.

Many of the names on the map are those of the crew of the voyage of reconnaissance, who were looking to divide among themselves and their neighbors a promising river valley, uninhabited except for a few Indians, who did not at the moment seem to present any serious obstacle to the plan. Through the names, some definite information on this abortive colony has been brought to light as the result of some ingenious work in the manuscript, Massachusetts legal records, by an architectural historian who is also a genealogical expert.[12] All that was known before of this voyage was the content of John Locke's copy of the map, plus a few contradictory and confused contemporary references. There is still a great deal not known about the project, including the relationship of these New Englanders to other groups in London and in Barbados, who were also interested in the Cape Fear area. Another voyage of reconnaissance from Barbados in 1664 resulted in another map and another better documented attempt at settlement, which lasted a few years but failed to become permanent.[13]

Maps showing the location of colonies after they were begun, or at least definitely projected, were more likely to find their way into print than the actual maps of reconnaissance. This kind of map was frequently used for the practical purpose of stimulating the interest of prospective settlers and investors. Such maps provided information for the public, although sometimes it was very sketchy information. One early instance was a map of Newfoundland published in 1625 (fig. 5), showing the location

[11] See E. G. R. Taylor, *The Mathematical Practitioners of Tudor and Stuart England, 1485-1714* (Cambridge, 1954).

[12] Louise Hall, "New Englanders at Sea: Cape Fear before the Royal Charter," in *The New England Historical and Genealogical Register*, 102, 2 (April 1970), 88-108.

[13] Cumming, *The Southeast in Early Maps*, no. 78. The attribution to Lancaster is no longer considered valid, the copyist being an anonymous member of the Thames School.

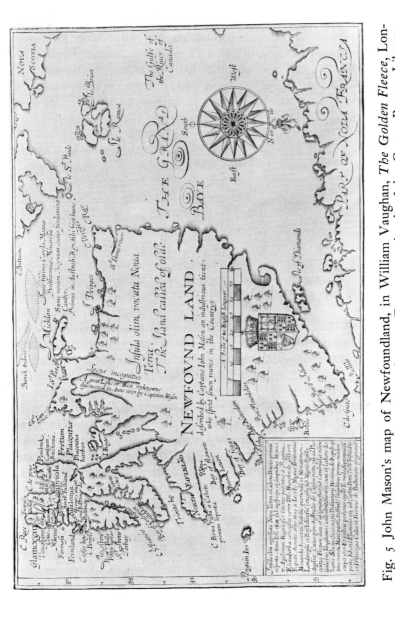

Fig. 5 John Mason's map of Newfoundland, in William Vaughan, *The Golden Fleece*, London, 1626, oriented with south at the top. From copy in the John Carter Brown Library. Reduced.

of English settlements on the Avalon Peninsula, including that of Sir George Calvert, who was later to become discouraged by the climate and transfer his colonizing activities to Maryland. The author of the map, Captain John Mason, who was later active in the settlement of New Hampshire, is described as "an industrious Gent: who spent seven yeares in the Countrey." Mason wrote of making a map in a letter from Newfoundland in 1617, the writing of which had been delayed by his exploring activities:

as huswives have many letts to good house wifry, frontletts, brace-lettes, partletts &c.—so have inletts, outletts, bayes, coves, &c. through their discovery been so many obstacles and hinderances to my duty, devourers of tyme, not affoording me leisure to thinck of writing, the which once effected I shall afford you a mapp thereof with a particuler relacion of their severall parts, natures, and qualities. I am now a setting my foote into that path where I ended last to discover to the westward of this land, and for 2 months absence I have fitted myselfe with a small new gally of 15 tonnes and to rowe with 14 oares (having lost our former) we shall visite the naturalls of the country with whom I purpose to trade....[14]

Mason's description of Newfoundland was printed in 1620 without his map.[15] Five years later, however, the poet-colonizer William Vaughan used it to illustrate his *Cambrensium Caroleia*, and again in 1626 it appeared in his *The Golden Fleece*, to which Mason contributed an approving poem in which he mentions spending "Winters sixe" in "Our New-found Ile."[16] The original manuscript of the map does not exist, and it is not known how closely it resembled the printed production, but certainly the Avalon Peninsula area shows the "30 as good Harbours as the world affords," which he claimed to have explored. Vaughan made use of Mason's map in order to show the locations of the new and struggling English settlements for which at the time he entertained high hopes, expressed in both prose and poetry heavily weighted with classical imagery and conceits.

Although Mason's map is the earliest English map of New-foundland which has survived, there may have been at least one other in existence about the same time. Richard Whitbourne, an

[14] J. W. Dean, ed., *Capt. John Mason, the Founder of New Hampshire* (Boston, The Prince Society, 1887), p. 220.
[15] *A Briefe Discourse of the New-found-land* (Edinburgh, 1620).
[16] Leaf b3v.

experienced mariner with many years of familiarity with the coast of the island, gave detailed descriptions in his books that must have been based on a map.[17] His latitudes are different from Mason's, and his place names differ enough so that he must have been using a different map. This lost map may have been one of his own making.

Another locational map from later in the century was compiled by John Thornton in London and engraved by him in 1673 (fig. 6). Into a general map of North America he introduced information from the first voyage sent out by members of the Hudson's Bay Company in the *Nonsuch* in 1668 before the company received its charter. A legend just east of James Bay reads, "at Prince Ruperts River Capt. Zachariah Gillam wintered who found a civil entertainment by the Natives being very willing to Traffique with the English their chief trade being bever skins." The map shows Rupert's River, named for Prince Rupert, who was to be the chief sponsor of the company, as well as Charles Fort, the name given to the buildings the crew of the *Nonsuch* constructed for their first "wintering over." This locational map was advertised in a newspaper[18] and seems to have been intended for public information. No actual settlers were solicited by the Hudson's Bay Company, whose members had no desire to promote settlement on the chilly shores of the Bay except for their trading posts, which they intended to keep under strict control.

Locational maps were, however, frequently used throughout the century in connection with the promotion of colonies. The Carolina Proprietors, for instance, made considerable use of such maps in their pamphlet literature.[19] In 1677 William Penn combined a map and broadside description in a single attractive publication advertising his projected settlement of West New Jersey, four years before he obtained his patent for Pennsylvania. The map was made from a plate that had already been published twice before, but with numerous revisions and the addition of side panels which considerably increased its size. The text of the description also was a revised form of a piece already printed. One of Penn's main purposes was to make clear the ease of access

17 *A Discourse and Discovery of New-found-land* (London, 1620), pp. 1–6; *A Discourse Containing a Loving Invitation* (London, 1622), pp. 3–12.

18 *London Gazette*, 20–24 February 1672/1673. Full-size reproduction in *The Blathwayt Atlas*, vol. 1, map 5 (Providence, 1970).

19 Cumming, *The Southeast in Early Maps*, passim.

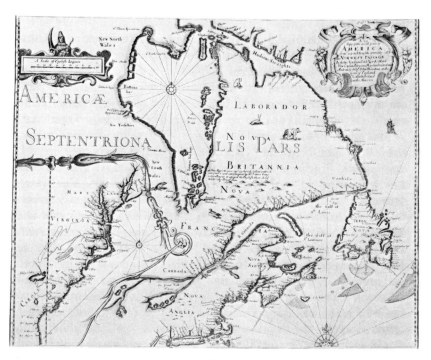

Fig. 6. John Thornton, *A New Mapp of the north part of America* [London, 1673]. From copy in the Blathwayt Atlas (Map 5), in the John Carter Brown Library. Reduced.

to the new colony, and the map emphasizes New York especially, with a newly engraved version of the harbor and its soundings inserted on the old map plate, while at the end of the description it is said that "to New-York, New-Jersey, Mary-land, and other Neighbouring Places, Ships will be going out from March until October."[20]

There is no information as to how many of the maps in this form fell into the hands of interested colonists, but apparently William Penn thought well of the idea. Four years later in 1681, when he was promoting the beginning of settlement in the colony named for his family, Penn approved the publication of another map with a printed description attached, this time of Pennsylvania.[21] The description states that the reason for making the newly compiled map was partly "to correct the Errors of those Maps that have taken in any part of this Country; for finding each Map at difference with itself, the Scale with the Latitude, and one Map with another, it was thought necessary to rectifie those mistakes, by a more exact Map, which hath been performed with as much Truth, Care and Skill, as at present can be, leaving room for time, and better Experience, to correct, and compleat it."

The southern boundary of Penn's colony was the 40th parallel, which on his map is placed well south of the present location of Baltimore, implying that his grant included a great deal of territory already settled by the Calverts of Maryland. The parallel is placed more than half a degree south of where it was on the New Jersey map sponsored by Penn four years earlier. Penn, or someone working for him, thus fired off a geographical gun that signaled the beginning of disputes that lasted far into the eighteenth century until the troubled boundary line was surveyed by Messrs. Mason and Dixon.

Maps that were intended to locate colonies sometimes showed wishful thinking of another kind, as did the map of Virginia and

[20] *A Mapp of New Jersey in America. By John Seller and William Fisher,* with text below in 4 columns, dated London, 1677. Map measures 42.5 cm x 92.2 cm. *The Blathwayt Atlas,* vol. 1, map 13. A separate facsimile of this map in color was published by the John Carter Brown Library, Brown University (Providence, 1957).

[21] *A Map of Some of the South and east bounds of Pennsylvania in America, being partly Inhabited* (London: John Thornton and John Seller, [1681]). Map measures 42.3 cm x 55 cm, with text below in 4 columns. *The Blathwayt Atlas,* vol. 1, map 15. A separate colored facsimile of this map was published by the John Carter Brown Library, Brown University (Providence, 1943).

the surrounding parts of North America made by John and Virginia Ferrar, first published in 1651.[22] In it is illustrated the concept that the journey from the Virginia settlements to the Pacific Ocean was short and easy, in the hope that Sir Francis Drake's New Albion, shown on the map, could be made a steppingstone, or a stepping-off place, for trading voyages to the Far East, where silks could be bought rather than raised from the silkworm in America. The Ferrar family were not specialist cartographers. They were known for their religious fervor, their book learning, their interest in handcrafts, and their early investments in the Virginia Company. By mid-century the company was no longer in existence, but their interest in the colony continued. By that time many people believed that the North American continent was considerably wider than could be covered in a ten days' journey on foot, as stated in a legend on the map, but this had not actually been established at the time. It was to be another twenty years before Virginians would go beyond the Appalachian Mountains and visually confirm the rumors of unending continental land west of their settlements on the tidewater and up to the fall line. Until the longitude of the west coast could be reliably established, there was room for hope, and the Ferrar map is an expression of that hope. It is useful historically for other reasons. It is one of the very few extant records of the Plowden patent for an area on the Delaware, called New Albion, which is recorded on this map with several pertinent place names. The name connects it with Drake's New Albion, giving it a meaning that was readily evident to Englishmen of the seventeenth century.

Once the settlement of any colony was begun, the maps of reconnaissance and of location were no longer sufficient. There was demand for maps of the colonies themselves, to show land ownership, boundaries, and access to harbors and towns. Where there was demand there was likelihood that a map would be made, but there was no certainty. Circumstances often prevented for many years the making of a good and useful map of a colony. One factor that seems to have made for delay was that a map made on the ground was necessarily dependent on the presence of a local surveyor who was qualified to produce a map. There are numerous bits of evidence that lead us to suspect that many

[22] *A Map of Virginia discovered to ye Falls . . . John Farrer Esq Collegit . . .* ([London], 1651). Cumming, *The Southeast in Early Maps*, no. 47 and pl. 29. *The Blathwayt Atlas*, vol. 1, map 22, is the same map in its fourth state.

of the surveyors in the new colonies were not highly qualified either by experience or by education. Of course all the colonies had surveyors for the routine job of measuring out the land that was to belong to individuals and to mark the bounds. Sometimes the marks were stones or cairns, but when possible they used trees, this being the easiest method in the short run. One of the few intercolonial boundaries actually surveyed in the seventeenth century, the line of approximately twenty miles separating Virginia and Maryland on the Eastern Shore peninsula, is shown as a double line of trees on Augustine Herrman's map of 1673. Trees were not always satisfactory marks, however. In 1681 in Burlington, New Jersey, one Thomas Harrison wrote to his brother in London saying, "Pray send me a Brass-Compass, with a Dyal for this Latitude; whereby the more easie to find the marked trees of my own Land, from other Men's." Mr. Harrison is said to have been in the colony only a year and a half and "being a poor Man, had not any Servants."[23] Undaunted by his situation, he was apparently making a single-handed attempt to correct the imperfect work of a New Jersey surveyor. It seems likely his problem was not unusual.

There is also some evidence in John Love's textbook on surveying, *Geodaesia*, published in London in 1688. Love had been a surveyor in the colonies before returning to London to become a teacher of mathematics and surveying. His book included a chapter entitled "Of Laying out New Lands, very useful for the Surveyors, in his Majesty's Plantations in America." In the preface he says,

I have seen Young men . . . often nonplus'd so, that their Books would not help them forward, particularly in Carolina, about Laying out Lands, when a certain quantity of Acres has been given to be laid out five or six times as broad as long. This I know is to be laught at by a Mathematician; yet to such as have no more of this Learning, than to know how to Measure a Field, it seems a Difficult Question. . . .

The sort of surveyor Love had in mind was hardly capable of making a map of a colony. Perhaps also some of the surveyors were handicapped in that their early experience had been in parts of England where the countryside was comparatively neat and

23 *An Abstract, or Abbreviation of some Few of the . . . Testimonys from the Inhabitants of New-Jersey* (London, 1691), p. 21.

uncomplicated. Unless they had worked in the Scottish Highlands they would not have been well prepared for the problems
they were to encounter in some parts of America, problems of
terrain and also those arising from the presence of a disapproving
native population.

Not all the surveyors, however, were routine people. Some of
them were very able and became important men in the new
communities. It should not surprise us, however, that complete
maps of most of the colonies were not made immediately. As was
to be expected, the smaller islands were the first to be surveyed.
There the matter of boundaries had been settled by nature, and
it was not impossible for one man, with a minimum of help, to
go over the whole territory himself and know it intimately.
Much greater problems arose, of course, with a large island like
Jamaica, where there was a great deal of mountainous terrain,
where some parts were practically inaccessible, and where a
native population could discourage the activities of surveyors.
On the mainland distances were even greater, and the wilderness
was a barrier not conquered, let alone mapped, until well after
the end of the seventeenth century.

The first surveyed map of an English-American colony was
drafted in 1617, and the subject was Bermuda. In this case there
were all the requirements for success. First, the Bermuda Company wanted and expressed willingness to pay for a survey and
a map. Second, the colony was a small cluster of islands, comprising less than twenty square miles, with no hinterland and no
native population. Third, and most important, Richard Norwood, a young man of twenty-four, who had recently come to
the islands in the hope of making his fortune diving for pearls,
was available for the job. More is known of him than of most of
our early colonial surveyors and cartographers because after returning to London in 1617 he went on to become a well-known
mathematician and produced a number of books. He even left a
sort of autobiography, although unfortunately it deals more fully
with his inner spiritual life than it does with his surveying.[24] The
original manuscript of his map does not survive, but it was en-
garved in 1626 by Abraham Goos—a Dutch engraver working

[24] The Journal of Richard Norwood, Surveyor of Bermuda, with introduc-
tions by Wesley Frank Craven and Walter B. Hayward (New York, 1945);
"Bibliography of Norwood's Writings" by William A. Jackson, pp. lix–lxiv.

in Amsterdam—and used by John Speed, of London, in an atlas of 1627. On the reverse side Speed printed Norwood's account of the early years of the settlement. Frequently reprinted and reengraved, the map became one of the best known of any American area in the seventeenth century. Later in his life Norwood went back to Bermuda and lived out his old age there. In the 1660s he was asked to revise and bring up to date his original survey, and so he made another map, accompanied by a book detailing the ownership of all the parcels of land. This second survey was not published at the time, but one of Norwood's two original manuscript copies has survived and is appropriately in the Archives of Bermuda at Hamilton, although it is in very poor condition. A copy made in 1678 and signed by Thomas Clarke, otherwise unknown, has coloring and decorative features in common with maps made by the Thames School mapmakers.[25]

The Lords of Trade as well as their predecessors, the Councils for Trade and Plantations, organized after the Restoration of Charles II, were instrumental in reminding the governors of the various colonies that maps were wanted and needed and that it was one of their obligations to arrange to have them made. The success of this prodding from London was variable. Sometimes the governors made great efforts to comply, and sometimes they ignored the requests. One of the better results was a map of Barbados by Richard Forde sent to the Lords of Trade by the governor of Barbados in 1680. There is reason to believe that Forde produced his map with little or no encouragement from the governor and that it had already been printed, possibly before 1675. Little is known of Forde except for the fact that he was a Quaker, and no other works of his have survived except for this map. One letter by him exists in a collection in the Boston Public Library, written to another member of the Society of Friends in Philadelphia, but it tells nothing of his work or of himself except that he lived in Bridgetown and that his handwriting was excellent. Probably, therefore, the orginal manuscript of the map, which does not survive, was a neat and attractive job. It was also a very intelligent work, full of information concerning roads, the locations of plantations with owners' names, the locations and type of sugar mills and whether they were wind-driven, powered by horses and oxen, or run by water power. This sort

25 *The Blathwayt Atlas*, vol. 1, map 24.

of information in convenient form was not only of use to the inhabitants of Barbados and to the administrators in London, but it is also the kind of information historians are constantly hoping for but seldom find in maps of the seventeenth century.[26]

It may well be asked why there were no attempts at cooperative surveys to produce maps of entire colonies. Conditions were apparently still too primitive for such an arrangement to be effective. Even many years after the colonies became independent, this sort of cooperative map was seldom attempted. This, however, was not entirely unknown in the seventeenth century. In 1671 the Council of Jamaica ordered a map of the island, providing that each of the district surveyors be responsible for his own area. Nothing more is heard of the proposal, however, and it may have been recognized as visionary. Within months the project was initiated all over again and entrusted to two men, only one of whom was a regular island surveyor. The map resulting was an advance over previous maps, but it was far from the product the council seems to have had in mind.[27]

The practical nature of the earliest maps of the colonies is apparent in the maps of rivers which were produced and which in many cases served the same purpose as maps of colonies, since the mainland settlements naturally clustered along rivers, the only good lines of transportation. For instance, a map of the Surinam River and its tributaries was, for all practical purposes, a map of the English colony that flourished there briefly until 1667, the date of the map.[28] The boundaries were little more than a legal fiction, unexplored and certainly unsurveyed. The same concept was applied elsewhere. For instance, in 1684 the governor of East New Jersey wrote to the proprietors saying, "An exact Mapp of the Countrey is not yet drawn, nor can you quickly expect it, for it will take up a great deale of time, charge and pains to do it."[29] In the same year a surveyor, John Reid, was sent to the colony, and four years later his map was engraved in

[26] Richard S. Dunn, "The Barbados Census of 1680. Profile of the Richest Colony in English America," in *The William and Mary Quarterly*, 3 ser., 26, 1 (January 1969), 3–30, makes use of this map.

[27] See discussion in volume 2 of *The Blathwayt Atlas* (Providence, 1975), pp. 189–190.

[28] *A Discription of the Coleny of Surranam in Guiana Drawne in the Yeare 1667*, manuscript map in ink and colors, unsigned, in the style of the Thames School, in the John Carter Brown Library, is reproduced in Dr. Ir. Cornelis Koeman, *Maps and Charts of Surinam. Links with the past* (Amsterdam, 1973).

[29] Samuel Smith, *The History of the Colony of Nova-Caesaria, or New-Jersey*, (Burlington, 1765), p. 186.

London by one "R. Simson." Reid's *Mapp of the Rariton River*[30] was for most purposes a map of the East Jersey colony, although no attempt was made to show any boundaries (fig. 7). Relatively few boundary maps were made in the seventeenth century. It was chiefly after 1700 that controversies over boundaries resulted in large numbers of maps.

Another river map, published about 1695 by John Thornton, was the *New Map of the Cheif Rivers, Bayes, Creeks, Harbours, and Settlements, in South Carolina* (fig. 8).[31] Remarkable for its detail, which can only be hinted at in a reduced reproduction, the map locates more than 250 settlements and plantations along the rivers. It was based on a magnificent manuscript by Maurice Mathews, a very able surveyor who had been a resident of Charleston since its first settlement. The publication of this map in London by Thornton and Morden, dedicated to the Proprietors of Carolina, suggests that the colonizers in London as well as the settlers themselves considered maps of rivers the practical equivalent of maps of their colonies, especially on the mainland where the boundaries were distant and unsurveyed.

The English Empire in America came slowly into being during the seventeenth century. It consisted of many heterogeneous units, and each one had its own cartographical development. The pace of this development varied greatly from one colony to another, but there was everywhere a demand for maps of practical usefulness which was often met satisfactorily. The maps mentioned in this paper illustrate various types, but other examples of all kinds would have served as well to indicate that there was an overall development that paralleled, and to some extent helped to stimulate, the independent English publication of maps. It should be clear also that there is a great deal still to be found out about the beginnings of mapping in the colonies. Not only are there maps to be discovered, but there are cartographers to be identified and studied, and there is a forest of archival material, published and unpublished, that has not yet been used by the historians of cartography. There is much still to be done before we can come close to knowing the whole story of the early mapping of the English colonies in America.

[30] Mentioned in Leo Bagrow, *History of Cartography* (Cambridge, 1964), p. 193, as printed in America, but it has long been known it was published in England. See Lawrence C. Wroth, *The Colonial Printer* (Portland, Me., 1938), pp. 284, 328–329. The name of the engraver, however, has not been found on any other contemporary production.

[31] Cumming, *The Southeast in Early Maps*, no. 118, plate 42, p. 36.

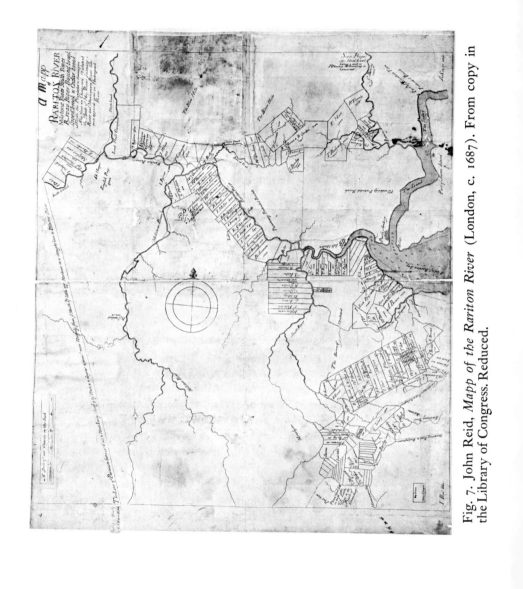

Fig. 7. John Reid, *Mapp of the Rariton River* (London, c. 1687). From copy in the Library of Congress. Reduced.

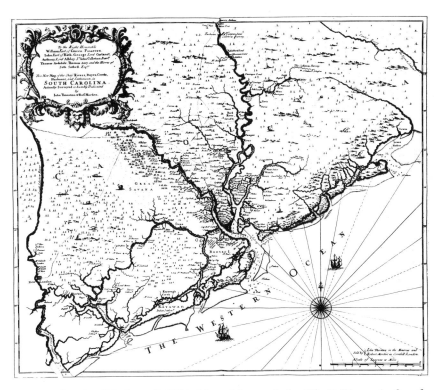

Fig. 8. [Maurice Mathews], *This New Mapp of the Cheif Rivers in South Carolina* (London, c. 1695). From copy in the John Carter Brown Library. Reduced.

IV

JOHN SELLER AND THE CHART TRADE
IN SEVENTEENTH-CENTURY ENGLAND

Coolie Verner

Until the end of the seventeenth century, English mariners depended on Dutch publishers for their charts and sailing directions, even of the coasts of England itself. In the beginning of
the century, a line of plattmakers had developed in a master-
apprentice relationship under the aegis of the Worshipful Company of Drapers.[1] The platts produced by the members of the
Drapers' School were portolan-type charts that were often copied
from Dutch originals and generally remained in manuscript.
Near the end of the century, John Thornton, of the Drapers'
School, began to published his own charts, and thus the manuscript map gave way to the printed chart.

Very few original surveys were made in England during the
seventeenth century, in the last quarter of which Pepys could
record that surveys of the English coast by Englishmen . . .
amounted to no more than:

1. The Sand-plat and Lynne-water, by the Trinity House.
2. Burlington Bay, by Captain Wood.
3. Isle of Wight and Portsmouth, by Lord Sandwich.
4. Jersey and Gernsey, by Gunman and Phillips about 1680.
5. The land-map to the Sand-plat from Deal to the Thames on Kent-

---

[1] Tony Campbell, "The Drapers' Company and Its School of Seventeenth
Century Chart-makers," in Helen Wallis and Sarah Tyacke, eds., *My Head is
a Map: A Festschrift for R. V. Tooley* (London, 1973), pp. 81–106. The relationship to the Drapers' Company was discovered by Professor T. R. Smith who
identifies this group as the "Thames School" but the designation "Drapers'
School" is preferred here.

side, and from the Thames to Yarmouth on the other, by Sir Jonas Moore.[2]

An instrument maker, John Seller,[3] made the first sustained attempt in England to compete with Dutch publishers, but in so doing he plagiarized Dutch material. Seller initiated a marine atlas that ultimately consisted of six volumes under the general title *The English Pilot*, with a volume each on the *Northern, Southern, Mediterranean, Oriental, West India*, and *African* navigation. However, Seller himself was responsible for producing only the first two volumes. In the 1670s and 1680s he also published a *Description of the Sands . . .* (fig. 1), *Coasting Pilot . . .* (figs. 2, 3), an *Atlas Maritimus . . .*, and an *Atlas Terrestris. . . .* However, the first wholly English marine atlas was *Great Britain's Coasting Pilot . . .* published in 1693. This was the work of Captain Greenvile Collins who between 1681 and 1688 surveyed the English coasts and produced 120 charts of which 49 were engraved and printed.[4]

Few intact copies of Seller's first publications are extant, and hence they are little known. Any analysis of them must necessarily include all items simultaneously, and is complicated by a number of factors:

1. Seller did not produce identical copies; consequently, it is virtually impossible to describe a complete copy.
2. His chart plates were printed in several different forms and used indiscriminately; thus, no specific sequence of printing and use can be clearly identified.
3. Large stacks of charts and text were printed and gathered at different times; hence, it is nearly impossible to order the sequence of these copies.
4. His earliest publications are scarce; therefore an insufficient number of copies exists to permit valid generalizations about the contents.
5. Latter-day booksellers have cannibalized copies, providing no assurance that any one copy is in its original form.

Despite these factors, this study attempts to analyze John Seller's earliest marine publications and their relationship to each other.

[2] J. R. Tanner, ed., *Samuel Pepys's Naval Minutes,* Publications of the Navy Records Society, 60 (London, 1926), 135.

[3] The name Seller is found in a number of spellings. His father was recorded as Sellers and John himself signed his name in that form frequently. On his charts and books the name is usually spelled Seller, so that form is adopted here. It has also been found as Sollers, Sallers, and Sallows.

[4] Coolie Verner, *Captain Collins' Coasting Pilot*, The Map Collectors' Circle, no. 58 (London, 1969).

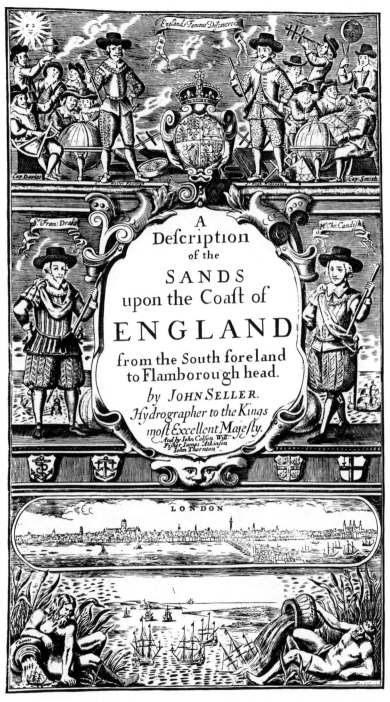

Fig. 1 Title page of *A Description of the Sands . . .* by John Seller
c. 1678.

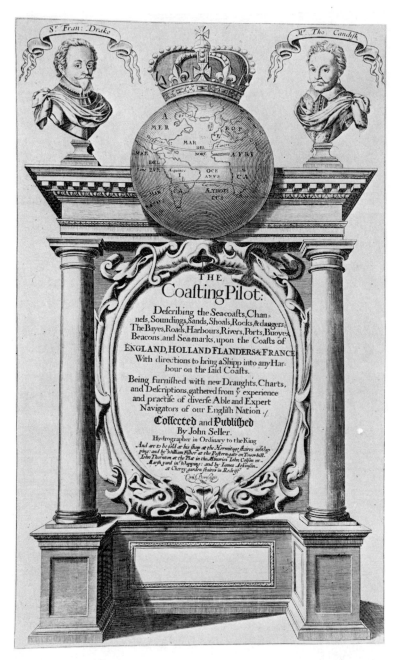

Fig. 2. Title page of the *Coasting Pilot* by John Seller, 1672.

Fig. 3. Title page of *The Coasting Pilot* (inset) c. 1678 by John Seller

JOHN SELLER

John Seller was an instrument maker at the sign of the "Mariner's Compass" near the Hermitage Stairs in Wapping. This area, on the north bank of the Thames east of the Tower of London, was the center for much of the marine trade during the seventeenth century. To it were attracted teachers of navigation, chartsellers, nautical publishers, and other similar artisans who depended on marine commerce. It was in this setting that Seller had his home and shop where, in addition to making and selling instruments, he sold charts and books.

Much has been written about John Seller, but most of it is more fancy than fact. He was born between 1627 and 1630, the son of "Henry Sellers, a cord wayner" of Wapping. On September 4, 1644, he was apprenticed to Edward Lowe of Whitechapel for eight years.[5] Little is known about Lowe except that he was admitted to the Freedom of the Merchant Taylors' Company "in virtue of his service to Robert Draper" on January 16, 1614. Presumably he was an instrument maker since that was the trade followed by his apprentice. The Merchant Taylors' was one of the older City Livery Companies and had ceased to consist only of tailors by the late Middle Ages.[6]

Seller completed his apprenticeship and became a Freeman in the Merchant Taylors' on October 25, 1654 and received the livery on November 10, 1676. Although he listed himself as a Merchant Taylor in his will Seller appears not to have been active in his company. He left his wife, "my five pounds for corne money in the Company of Merchant Taylors . . ." and his youngest son Jeremiah became a member by patrimony on September 1, 1703.

Seller was taken into the Clockmakers' Company in 1667 and served as warden from 1692 to 1696.[7] The Clockmakers' were more in Seller's line, and he appears to have been more active in this company as it more closely fulfilled his needs than did the Merchant Taylors. It was not customary for an individual to hold membership in more than one company, but until 1667 the

[5] From the records of the Merchant Taylors' Company in London.

[6] The information about Lowe is by courtesy of M. W. B. Sanderson of the National Maritime Museum, Greenwich.

[7] F. J. Britten, *Old Clocks and Watches and Their Makers* (London, 1904), p. 691. This is ample reason to doubt that the John Seller listed by Britten as admitted to the Clockmakers in 1667 was actually the same person. Professor E. G. R. Taylor states that they were.

instrument makers did not have affiliation with a particular company.

Following his apprenticeship, Seller lived for a time in the parish of St. Katherine-by-the-Tower. He and his wife, Elizabeth, apparently had two daughters and a son during this period.[8] Little is known of the two daughters, Elizabeth and Thomasin, but later, John Junior established himself as a bookseller with money borrowed from his father. His first shop was "The Star Next to Mercer's Chapel in Cheapside," and in 1688 he was listed at the "West End of St. Paul's Church yeard."[9] About this time, John Seller identified himself as Senior which was indicated on one of his engraved title insets.[10] John Junior outlived his father by only slightly more than a year and was buried at St. John's, Wapping, on October 16, 1698.

A fourth child, Jeremiah Seller was the youngest and longest surviving son. It is doubtful that he was born in the parish of St. Katherine because he does not appear in the records of the business until a year after his father's death.

Among those employed by John Seller was Charles Price, an instrument maker, who had been apprenticed to George Steven in the Clockmakers' Company. He entered into partnership with Jeremiah Seller in 1699 and they produced instruments for the navy as well as publishing maps. They issued a new volume of *The English Pilot* on Africa which the elder Seller may have been working on at the time of his death.[11] They also published a new edition of *The English Pilot, West India* with new charts constructed by Price.[12] However, they were not successful and their plates and stock were sold to Mount and Page in 1705. By 1706 Price had joined John Senex to produce atlases, which were superior to those of John Seller, and earned for them a comparable posterity.

Apparently John Seller was a Baptist at a time when Noncon-

[8] Parish records, St. Katherine-by-the-Tower.

[9] These addresses are from advertisements in the *London Gazette* 30 August 1686, 16 May 1687, and 28 June 1688.

[10] Coolie Verner, "Engraved title-plates for the folio atlases of John Seller," in Wallis and Tyacke, eds., *My Head is a Map*, John Seller, Junior, was admitted to the Stationers' Company on 26 March 1686. This may be the date when his father began to identify himself as "Senior."

[11] Coolie Verner, "Bibliographical Note" in the facsimile edition of *The English Pilot . . . Africa* (Amsterdam, 1973).

[12] Coolie Verner, *A Carto-Bibliographical Study of The English Pilot. The Fourth Book . . .* (Charlottesville, 1960).

formists were especially unpopular in England.[13] The post-Restoration period was marked by treasonable acts and plots by such groups as the Fifth Monarchy Men who drew support from Nonconformist groups.[14] One plot was led by Ensign Thomas Tonge in 1662. Six men were arrested for high treason including Tonge, George Phillips, Francis Stubbs, James Hind, Nathaniel Gibbs, and John Seller. The trial was held in "Justice Hall in the Old-Bailey, London" on December 11, 1662.

In the published report of the trial, it is clear that Seller was not actually involved in the plot, although he was a friend of the conspirators.[15] He was arrested on the testimony of a government agent who saw him talking to another defendant. On his own behalf Seller said:

Here is that, that I am accused of, That I delivered Arms, which is altogether false; My Lord Major can bear witness, *Wapshot* confessed, he told me such a thing. I did ever abhor any such thing. . . .

In spite of the testimony that clearly indicated his innocence of the charge, Seller was found guilty along with the others and sentenced.[16]

Seller and Hind escaped execution, although the others were hanged on December 22, 1662.[17] Seller remained in Newgate Prison until the spring of 1663 when he was allowed one month's release on bail by an order dated April 21.[18] Before the month was over he petitioned the duke of York that "being bound in 3,000 bayle" for his surrendering himself "to that doleful Prison of Newgate within four days . . ." he sought "an order for his longer inlargem't . . ." while he sought a pardon.[19] This extension was granted by an order issued May 19, 1633.[20] In time a pardon was granted but it was not confirmed because Seller could not produce the usual fee; he therefore petitioned directly to the king:

[13] S. P. Dom. Car. II, 321, numbers 300 and 305 are records of Seller securing license for Baptists in June 1672.

[14] B. S. Capp, *The Fifth Monarchy Men* (London, 1972), p. 209.

[15] William Hill, *A Brief Narrative of that Stupendous Tragedy* . . . (London, 1662), p. 55. See also *True and Exact Relation of the Arraignment* . . . (Wing 2444A).

[16] Ibid., p. 61.

[17] Cal. S. P. Dom. Car. II, 1661–1662, p. 602.

[18] PRO, S. P. 44/9.

[19] Ibid., 29/72.

[20] Ibid., 44/9.

To the KINGS most Excellent Majestye

The humble Petition of John Seller Compasmaker Humbly Sheweth,

That your Petitioner hath a Deep Sence of Your Majestyes boundless Mercy bieng Gratiously Pleas'd to pass your Grant for your Petitioners Pardon, In all Submission Presents Your Majestye his Impoverisht Condicion by Reason of his long Restraynt and Imprisonment, being therby Reduc't to Extraordinary Exigencyes, for the mayntenance of his Wife and foure Small Children, and haveing thereby Contracted some Debts for there Support, which Your Majestyes humble Petitioner is not in a Present Capacitye able to Discharge, And your Petitioner being still under Conviction of that horrid Cryme which his Soule abhors, lives a languishing Dying Life, until he be Absolv'd there of by Confirmation of Your Ma.[ties] Gratious Pardon,

Your Petitioner & Obedyent Subject doth most Humbly Pray and Implore Your Majestyes Extended Mercy in further Comiseration of such his Indigent Condicion, To Remitt the accustomed Fees for his Pardon. . . .[21]

The first notice of Seller's activities is found in the *Young Seaman's Guide* by Timothy Gadbury published in 1659. This contained "a notice that all the necessary instruments and charts could be had new or second-hand of John Seller the compass maker, at his Shop by the Hermitage Stairs, Wapping."[22] In 1672, he advertised in the *Term Catalogue* that he had for sale:

Mathematical Instruments, Meridian Compasses of all sorts and sizes in Brass and Wooden boxes, all sorts of running Glasses, Azimuth Compasses, Amplitude Variation Compasses, Semi-Circles, *Davies*-Quadrants, Cros[s]-staves for forward and backward Observations, *Gunter's* Cross-Staffs, *Gunter's* Bows, *Hood's* Bows, Ploughs, Removing Quadrants, Triangular Quadrants, Inclinatory Needles for finding the latitude of a place without Observation of Sun Moon or Stars, Universal Ring Dyals for the Hour of the Day and Latitude of the place, *Gunter's* Sectors, *Gunter's* Rules, plain Scales, Scimical Quadrants, Rules for Carpenters Gunners and other Artificers, Horizontal Dials and Pocket Sun-Dials for any Latitude, Gunner's Rules Heights and Callapers, Brass Compasses of all sorts, Dividers, Gaging Rods, plain Tables, Theodolites, Circumferenters, Loadstones, Globes with Sphears of the Heavens etc. Microscopes and Telescopes, Lamps, Perspective Glasses, Maps of the World in all sizes, and of any par-

[21] Ibid., 44/15.
[22] E. G. R. Taylor, *The Mathematical Practitioners* (Cambridge, 1954), p. 92.

ticular County; with any other Mathematical Instruments whatsoever.[23]

In addition to making and marketing instruments, Seller was
also interested in their use. In the March 1666 issue of the *Philosophical Transactions* (later of the Royal Society), a correspondent posed some "Magnetical Inquiries" which John Seller answered in the issue of June 1667.[24] He wrote a book that covered
many aspects of navigation and the relevant mathematics, including the use of instruments, an almanac of the moon, tables of
the sun's declination, and similar matters of interest to mariners.
This work, *Praxis Nautica or Practical Navigation* . . . was published in 1669 and advertised in the *Term Catalogue* in November as being for sale by Seller in Wapping and by John Wingfield
in Crutched Friers. This book was immediately popular and
established a reputation for Seller that led him deeper into publishing. It was reissued frequently in subsequent years.

On February 14, 1672, Seller wrote to the commissioners for
the navy asking for a contract to supply compasses and glasses.[25]
This was followed by a letter to Commissioner Thomas Middle-
ton at Chatham on February 20 "requesting an order for serving
half-hour and half-minute glasses into the stores there. . . ."[26] He
received such an order and continued to supply instruments for
many years. After John Seller's death, his widow Elizabeth received her own contract from the navy in 1698, and Jeremiah
Seller and Charles Price also had a contract. From 1698 to 1707
there are frequent references to payments made to Elizabeth
Seller and to Elizabeth and Jeremiah Seller for compasses.[27] At
one point Elizabeth complained that she was paid only £4 for
Azimuth Compasses while John England was being paid £5. The
Navy Board ruled in her favor.[28] Seller and Price did not produce satisfactory instruments and lost their contract with the
navy in 1707 as a result of complaints from ships' officers.[29]

[23] *Term Catalogue*, Hillary, 7 February 1672.
[24] *Philosophical Transactions* (later of the Royal Society), no. 26, 3 June
1667, p. 478. The original queries were in number 23, 11 March 1666, pp. 423–424.
These original queries may have been made by Sir Nicholas Millet.
[25] S. P. Dom. Car. II, 322, no. 116.
[26] Ibid., no. 137.
[27] I am indebted to Commander W. E. May for supplying this information
on the Sellers' relationships with the navy.
[28] PRO, Adm. 106/623, Navy Board Misc., in-letters "S."
[29] Ibid., 106/617, Navy Board Misc., in-letters "C."

Instruments were Seller's basic trade and undoubtedly provided his primary source of income, but he was not content to remain solely in that occupation. Subsequently, as noted, he became involved in publishing for the marine trade even to the extent of making charts, although he was not a recognized plattmaker. Two manuscript charts[30] made by him are known to exist but his competence in this regard prompted Samuel Pepys to note:

My lord is, upon this trial of ours at our coming into the Channel, mightily convinced and angry at Seller's platt (made on purpose for him this voyage) proving worse than the Master's old Dutch ones.[31]

During most of his life, John Seller maintained his principal establishment in Wapping but from time to time—usually during periods of prosperity or expectations of success—he also listed other addresses. These secondary shops can be identified with some precision through advertisements in either the *Term Catalogues* or the *London Gazette*.[32] By relating the address listed in an imprint to the time Seller is known to have maintained a secondary shop, it is possible to register his undated publications with reasonable accuracy.

To 1671   *Wapping*

Both the typographic title page and the engraved title inset for *The English Pilot, Northern*, 1671 have only the Wapping address. The typographic title page was printed before March 24, 1671 as it does not list Seller as "hydrographer," while the engraved title inset was cut after that date.

1671–1675   *And in Exchange Alley in Cornhill*
              And in Exchange Alley near the Royal Exchange

These two addresses are assumed to refer to the same location. This shop was listed in the *London Gazette* on March 6 to 9, 1671 and continued to be used until June 17, 1675 when it last appeared in the *Gazette*. During this time Seller prepared the engraved title insets for his *Atlas Maritimus, Atlas Terrestris,*

---

[30] Both charts are on vellum and are preserved in the Bibliothèque nationale in Paris. One is "A draught of Bombay" and the other an untitled chart of the west coast of India dated 1684.

[31] Edwin Chappell, ed., *The Tangier Papers of Samuel Pepys*, Publications of the Navy Records Society, 73 (London, 1935), p. 242.

[32] I am indebted to Sarah Tyacke for assistance on advertisements in the *London Gazette*. See her "Map-sellers and the London map trade c. 1650–1710," in Wallis and Tyacke, eds.. *My Head is a Map.*

and the *Coasting Pilot*, as well as the typographic title page for *The English Pilot, Southern*.

1675–1678   *Wapping*

By 1675, John Seller was experiencing financial difficulties and abandoned his second shop. From April 13, 1676 to February 12, 1677, Seller's advertisements listed Wapping with the added line "And John Hill in Exchange Alley." This suggests that Hill may have taken over Seller's original shop in Exchange Alley. In the *London Gazette*, October 1691, Hill is listed as "over-against Jonathan's Coffee House in Exchange Alley in Cornhill." Quite possibly Seller merely used Hill's address from 1671 to 1675 rather than having his own shop.

During this period, Seller printed the typographic title page for *The English Pilot, Oriental* and the *Atlas Maritimus*.

1675–1681   *And in Pope's Head Alley in Cornhill*

Seller attempted to reestablish himself with a new series of publications and opened this secondary shop that was listed in the *London Gazette* August 15, 1678 when he advertised his *Atlas Minimus*. It appeared for the last time in an advertisement for *Atlas Coelestis* . . . in the *Term Catalogue* May 16, 1681. The engraved title page for *A Pocket Book* . . . lists this address without date while that for *Atlas Minimus* had this address added to the imprint after the plate had been cut.[33]

1681–1682   *Wapping*

No advertisements appeared in this interval.

1682–1686   *And the West Side of the Royal Exchange*

The earliest use of this address is found on the typographic title page dated 1682 for the miniature edition of *Atlas Maritimus*. It appears in advertisements in the *London Gazette* from September 29, 1684 to August 30, 1686. During this time Seller engraved a new title inset for *The Coasting Pilot* (II) which he reissued about 1685 using remainder stock. He also made an engraved title inset for *Atlas Anglicanus*. . . .

1687–1690   *Wapping*

Only the Wapping address was used in advertisements in the *London Gazette* during this period.

1690   *And at his Shop in Westminster Hall*

This address appeared once in an advertisement in the *London Gazette* on May 26, 1690.

---

[33] In the *Term Catalogue* for May 1679, Seller lists his address as "in Cornhill near the Royal Exchange" and in November it is "Pope's Head Alley in Cornhill." It is assumed that these both refer to the same shop.

1690–1697 *Wapping*

Until his death in 1697, Seller used only his original Wapping address. It was continued by Seller and Price in their advertisements.

John Seller died of "dropsy" and was buried at St. John's, Wapping, in May 1697. His will had been signed May 19, 1697[34] and was witnessed by John Hodges, Francis Spridell, and Charles Price. In disposing of his estate, Seller specified "that twenty shillings shall be laid out in penny loaves and be distributed amongst the poor of the parishe of St. John Wapping. . . ." One third of his estate was left to his widow, Elizabeth, and the rest to Jeremiah Seller. In regard to his eldest son, John, he noted "I have already given and advanced to and with my eldest son John Seller the summe of the vallue of the summe of five hundred pounds of lawfull money of England . . . which is his full part or share. . . ."

Elizabeth Seller was named as sole executor of the will and it was entered for probate immediately. She survived her husband by fourteen years and was buried at St. John-at-Wapping on December 26, 1711.

PUBLISHING ACTIVITIES

After his pardon was confirmed and his release from Newgate Prison achieved, Seller apparently made a trip to Holland where he came upon a collection of some sixty-three old and worn copperplates that had been used to print the charts in a Dutch sea-atlas.[35] Seller immediately saw that these would enable him to produce his own sea-atlas at little expense in competition with the Dutch publishers who then dominated the market. He brought these plates home and began preparing his atlas.

The plates he acquired had been made originally in 1620 by Jan Jansson for a counterfeit edition of Blaeu's *Het Licht der Zeevaert.* . . . Jansson had used the plates for several editions of his copy of Blaeu and then they had passed to J. VanLoon who published an edition of *Le Nouveau Flambeau* . . . in 1650.[36]

---

34 PCC, *Pyne*, folio 103.

35 There is no positive evidence to support the hypothesis that Seller made such a trip but it seems improbable that he could have known of or acquired these plates in London. It is possible that they may have been shipped to London in a lot of scrap metal but this seems unlikely.

36 On Jansson's counterfeit, see R. A. Skelton, "Bibliographical Note," in the facsimile of Blaeu's *Het Licht der Zeevaert* . . . (Amsterdam, 1964). Skelton did not know of the VanLoon edition of 1650 when he wrote this note.

Some fifty-three of the plates had been used by VanLoon (fig. 4) and ten other plates came from various other sources. The plates included charts of northern waters, the coasts of England and Ireland; the Continent south to the Mediterranean; and northern Africa. Seller saw that these would enable him to prepare a series of volumes covering much of the world and, where necessary, he could supplement this collection with plates of his own.

Seller announced his intention to do this in a postscript to his *Praxis Nautica* of 1669:

I do hear make known unto you, that I intend, with the assistance of God, and am at present upon making (at my own cost and charge) a Sea Waggoner for the whole World, with Charts and Draughts of particular places, and a large Description of all the Roads, Harbours, and Havens, with the Dangers, Depths and Soundings in most parts of the World, which work was never yet performed by any. . . .[37]

As indicated in this announcement, Seller was "at present upon making . . ." his proposed Sea Waggoner. This indicates that he had the Dutch plates in hand and was preparing them for his use. The first volume known to have been issued was his *The English Pilot, Northern*, 1671, but there is sufficient reason to assume that he had in print the first volume of his proposed series at the time this prospectus was written. Such an earlier volume is not known to be extant, but there is a variety of evidence to support the hypothesis that it was in fact issued before the copies, dated 1671, which still survive.

Samuel Pepys noted the existence of an earlier edition when he wrote:

Mr. Phillips at sea examined and showed me how Seller's book in 1668 was the very same platts with the Dutch without a Dutch word so much as turned into English, much less anything in the maps altered.

And he says that he knows it to be true and Seller will not deny it, and that he bought the old worn Dutch copper plates for old copper, and had them refreshed in several places, and has used them in his pretended new book.[38]

From this comment, it is clear that Pepys and Phillips had a copy of the work and that they were examining an edition of *The*

---

[37] Quoted in Taylor, *The Mathematical Practitioners*, p. 380.
[38] Chappell, ed., *The Tangier Papers*, p. 107.

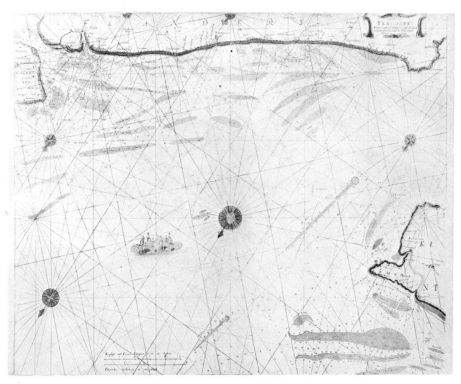

Fig. 4. A chart of Flanders by John Seller. This extension plate for Seller's chart of the Channel c. 1670 was originally a Dutch plate made by VanLoon for his edition of Jansson's *Le Nouveau Flambeau de la Mer*. Seller replaced the original Dutch title with his own and added the scales.

English Pilot, Northern earlier than that known to have been published in 1671. Furthermore, no book by Seller is known that was published before his *Praxis Nautica* in 1669. It is unlikely that Pepys would have made an error in the date of the volume being examined.

There is further supporting evidence for an edition of *The English Pilot, Northern* dated 1668 which can be inferred from existing copies of 1671. The collation and pagination of this edition is:

> 3 p.i., a–d², B–G², a–b², H–Z², Aa², Bb¹, Cc–Ii²,
> pp. [11 unnumbered leaves], 1–24, 1–8, 25–94, 97–124.

The first inserted leaves (a–d²) were printed after March 1671 and contain a *Privilege* and material that was in Seller's *Praxis Nautica* published in 1669. The second set of inserted leaves (a–b²) contains an *Appendix* . . . which is known in two forms. The three preliminary leaves include an engraved title plate probably made in 1671, a typographic title page dated 1671, a leaf with a dedication to James, duke of York, and a preface dated 10 November 1670 on the verso. This leaf and the title page may be conjugate but this could not be determined in any copies examined.

The missing signature Bb2 (pp. 95–96) is explained by Seller's note in his preface that "The Whole Work I have divided into four Books, and these into several Parts, that they may be had severally, for the ease and convenience of the Buyer." In the composition of the volumes, part one included B–G², part two H–Z², Aa², Bb¹, and part three Cc–Ii². This suggests that the parts were set and printed separately and since the text for part two was one leaf short of a full signature, that leaf, (Bb2), was omitted and an error made in the pagination.

The collation and pagination of *English Pilot, Northern*, 1668 would be the same as that of 1671 without the preliminary leaves and the interpolated material. Thus, it would read, 1 pl., B–Z², Aa², Bb¹, Cc–Ii², pp. 1 pl., 1–94, 97–124. Further differences between the 1668 and 1671 editions are discussed below with respect to the charts.

After publishing *English Pilot, Northern*, 1668, Seller acquired a copy of a manuscript chart of the Thames Estuary by Sir Jonas Moore. He had this engraved by James Clarke without permission. Pepys noted that Phillips:

says that Sir Jonas Moore himself told him that Seller did print a platt of his survey as if it had been his, before he himself had ever finished his draught of it. . . .[39]

The date of the Moore manuscript is not known. If Seller acquired his copy of it while *English Pilot, Northern, 1668* was in press, he may have used it with an appropriate text, such as an *Appendix* inserted into the sheets. This is suggested by an entry for an appendix in the index for the 1671 edition following page 24 and consisting of "p. 1, 2, 3, 4, 5, 6."[40] Undoubtedly, Seller received severe criticism for his plagiarism, for he quickly had a new plate engraved by Francis Lamb that acknowledged the chart to be by Moore (fig. 5).[41] This version was probably engraved by Lamb because Clarke was busy making other folio plates for Seller and could not produce this plate quickly enough to satisfy him.[42]

Seller prepared a text to accompany the Lamb plate and published his *Description* . . . in 1671 (fig. 6). This acknowledged Moore in the title. The text was then reset and used as an *Appendix* with the publication of the 1671 edition of *The English Pilot, Northern.*

On the strength of this work and the *Description*, Seller was granted a Royal Privilege dated March 22, 1671. This enjoined "any person to print any work, under other titles, reprinting, or counterfeiting, for thirty years . . ." and "the import from beyond seas of any such books or maps. . . ."[43] The complete text of this was set in black letter and included in the material added to *The English Pilot, Northern,* 1671 (a–d²). Additionally, Seller was designated "Hydrographer to the King" by a warrant issued March 24, 1671.[44] After that date, Seller added this designation to his map and title page imprints.

[39] Ibid., p. 108.

[40] It appears that this index was copied without correction from the 1668 edition as indicated by the maps.

[41] There is no sure evidence to indicate the order in which these plates were made but it seems illogical for Seller to have two nearly identical plates made and used simultaneously.

[42] Clarke engraved a number of plates for Seller at this time. No doubt Lamb was considerably more expensive than Clarke as he had a better reputation at the time.

[43] S. P. Dom. Car. II, 288, no. 103 and S. P. Dom. Entry Book, 34 f. 76.

[44] S. P. Dom. Entry Book, 34 f. 77.

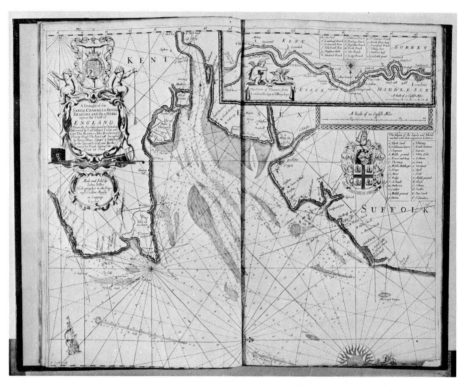

Fig. 5. A chart of the estuary of the River Thames by John Seller based on the survey by Sir Jonas Moore.

A

DESCRIPTION

OF THE

Sands, Shoals, Buoys, Beacons, Roads, Channels, Sea-Marks, on
the Coaſt of *E N G L A N D*, from the *S O U T H-*
F O R E L A N D to *O R F O R D-N E S S.*

SHEWING

Their Bearings and Diſtances from moſt of the Eminent Places on the
Land : VVith the depths of water on the Sands, as alſo in the Channels between. VVith ſeveral other things neceſſary to be known.

Being accommodated with a New and Exaƈt Draught of the Sands,
according to the ſaid Deſcriptions.

Diſcovered by {
Capt. GILBERT CRANE,
A N D
Capt. THOMAS BROWNE
} Elder Brethren of the *Trinity-Houſe.*

And the Sea-Coaſts Surveighed, by *Jonas Moor* Eſq; both by Warrant from His R O Y A L H I G H N E S S the Duke of *Y O R K.*

And by His ſpecial Licenſe and Approbation.

Publiſhed by J O H N S E L L E R.

L O N D O N,

Printed by *John Darby,* and are to be ſold by *John Seller,* at his Shop, at the ſign of the *Marinerr-*
Compaſs, at the *Hermitage-ſtairs* in *Wapping :* And by *John Wingfield,* Book-ſeller, right
againſt St, *Olaves* Church, in *Crutched-Fryars,* MDCLXXI.

Fig. 6. Title page of *A Description of the Sands,*
. . . by John Seller 1671. This work was an expansion of his earlier *Appendix.*

It is unlikely that Seller would have received either the privilege or the designation hydrographer only on the basis of the *Description*. Consequently, this seems to provide further evidence supporting the hypothesis of an edition of *The English Pilot, Northern* in 1668. Seller's predecessor with the title of hydrographer to the king was Joseph Moxon, who was a far more distinguished cartographer and nautical publisher. When Moxon was appointed hydrographer in 1662 it was on the strength of a petition to the King "signed by John Newton, D. D., and thirteen other individuals, mainly professors of mathematics."[45] These awards coupled with his bail and pardon, suggest that Seller had influence in court circles. Professor E. G. R. Taylor says that "he enjoyed the patronage of Sir Nicholas Millet of Battersea, a gentleman interested in the variation of the compass."[46] Seller dedicated the book to James, duke of York, who may have been the influential source. Pepys alluded to such influence when he reported "that the other plattmakers did article against Seller, but seeing his interest too strong for them, they were forced to sit down and be quiet."[47]

After receiving the privilege and title, Seller reissued *The English Pilot, Northern* in early summer 1671. The title page was in press or in print before March when he was named hydrographer as that does not appear after his name although it is present on the engraved title inset. In 1672, Seller published *The English Pilot, Southern, The Coasting Pilot*, and another edition of the *Description*. He may also have published other items, but these have not been identified. The privilege suggests that he issued a *Sea Atlas*, but this is not known.

With *The English Pilot, Northern* and *Southern* complete, Seller began to prepare the other volumes he planned to include in his proposed Sea Waggoner. In the preface to *The English Pilot, Northern*, 1670, he indicated that he planned to issue four books including *The English Pilot, Northern, Southern, Oriental*, and *West India*. The preparation and printing of the text for all the volumes began simultaneously, and Seller managed to complete twelve pages of *The English Pilot, West India*, twenty-four pages of *The English Pilot, Oriental*, and forty-nine pages

---

[45] Carey S. Bliss, *Some Aspects of Seventeenth Century English Printing with Special Reference to Joseph Moxon* (Los Angeles, 1965), p. 17.

[46] Taylor, *The Mathematical Practitioners*, p. 92. Taylor does not cite her sources so this matter could not be verified.

[47] Chappell, ed., *The Tangier Papers*, p. 108. Nothing has yet been discovered to explain this statement.

of *The English Pilot, Mediterranean*, which was not in his original plan. At the same time, he was attempting to make the necessary plates for the many maps that would be required.

Since John Seller was not a printer, he had to engage one to handle the letterpress text designed to accompany the maps in each of his publications. In selecting a printer, Seller chose a man of about the same age as himself who was also a Nonconformist —John Darby, who maintained a shop in London for many years and was succeeded in it by his son, John. Both father and son appeared frequently in the *Stationer's Register* where they entered their publications.

John Darby, who was born in Leicestershire, was the son of John Darby, a husbandman. The young Darby was apprenticed to John Hides for seven years on September 6, 1647 in the Stationers' Company. He did not become a Freeman of the company until November 6, 1666, and he received his livery on June 3, 1689. During the post-Restoration turmoil, Darby, like Seller, was involved with the law for publishing unlicensed books which was a less serious offence than was the treasonable act in which Seller had been implicated. On January 21, 1668, a warrant was issued for the arrest of Darby and his servant Gains.[48]

Darby printed unlicensed tracts for the Quakers and was confined to the Gatehouse, Westminster. In a petition seeking admission to the prison submitted later in January, his wife, Joan, claimed that:

he was not acquainted with the pernicious things in the book he printed, the copy being brought him by piecemeal, and the author William Penn, sometimes dictating to the compositor as he set the letters.[49]

Roger l'Estrange was involved in the investigation of unlicensed printing and on April 26, 1668, wrote to Joseph Williamson about Joan Darby as follows:

She'll confess nothing, for she and Brewster are a couple of the craftiest and most obstinate of the trade. I see nothing against Darby himself, so that only Brewster stands answerable, and printing does not concern her.[50]

Again, in early May 1668, Joan Darby petitioned "Lord Ar-

[48] S. P. Dom. Entry Book, 28 f. 12.
[49] Ibid., Car. II, 233, no. 140.
[50] Ibid., 239, no. 5 and no. 6.

lington to promote her petition to the king and council for liberation of her husband, who is much injured in health, and his life hazarded by his close confinement."[51] On May 7, a warrant was issued for the release of John Darby from the Gatehouse and the note that:

John Darby, printer, entered into a recognizance of 100 l. and 2 others of 50 l., for his personal appearance when required, and for his printing nothing contrary to the Act for printing in the meantime.[52]

John Darby printed most of Seller's publications issued between 1668 and 1675. His name appears in the imprint of the typographic title pages of the *Description, The English Pilot, Northern* and *Southern,* and the *Atlas Maritimus.* No doubt he also printed the *Coasting Pilot,* but it does not have a typographic title page. The relationship between Seller and Darby seems to have ended when a Combine took over and he does not appear to have printed any of Seller's small items issued after that date.

By 1676 or 1677, Seller had reached the limits of his financial resources and was forced to involve others in his project. In attempting to produce three volumes simultaneously, he was faced with printing costs for the text and the cost of engraving and printing the folio plates for the charts. The cost of preparing the required number of map plates was beyond the means of any but the most prosperous publisher.

Seller had an engraved title plate with insets for an *Atlas Maritimus* and an *Atlas Terrestris* which he could use to form a collection of appropriate maps on demand as was usual in the map trade at the time. For his *Atlas Terrestris* he printed several separate folios of text that could be assembled at any time. These included Asia, Africa, America, the Seventeen Provinces, and a description of the earth. These sheets of text would be gathered with the title plate and maps (including those of other publishers) in stock as appropriate.

In 1675, Seller prepared a typographic title page and twelve pages of text for his *Atlas Maritimus* and used his own maps, including some from the Dutch plates. In addition, he usually advertised some of his maps as they were ready for individual sale. Rather than concentrating exclusively on the plates required for his proposed volume of *The English Pilot,* Seller often wan-

[51] Ibid., no. 155.
[52] Ibid., Entry Book, 28 f. 1 and 28 f. 13.

dered afield with such items as the seventeen provinces of Germany or the city of Maestricht, but none of these ventures were productive enough to sustain him.

In the Preface to *The English Pilot, Mediterranean* published in 1677, Seller wrote:

And here I thought good publickly to advertise the Reader, that for the better Management of my so Chargeable and Difficult an Undertaking, I have accepted the Assistance of my worthy Friends, Mr. *William Fisher*, Mr. *John Thornton*, Mr. *John Colson*, and Mr. *James Atkinson* as my Copartners in the *English Pilot, Sea Atlas,* and in all Sea-*Charts,* Plain and Mercator; We resolving unanimously (by Divine Assistance) to spare neither Cost nor Pains, to render the whole of this Design the most Compleat of any Extant.

The partners in the Combine who took over Seller's project were an interesting group. William Fisher was a successful nautical publisher on Tower Hill whose firm ultimately became Mount and Page. In time, he acquired the entire series of maps and publications initiated by Seller which continued in print for most of the eighteenth century.[53] John Thornton was a plattmaker of the Drapers' School who maintained his shop in the Minories near Tower Hill and was far better equipped to produce maps than was John Seller. James Atkinson and John Colson were leading teachers of navigation in London. All of them were, in effect, competitors of Seller in one way or another.

Seller turned over to this Combine all of his plates and publications. As the plates were used, the imprints were altered to show ownership by the Combine, and the order of the names in the imprint indicates the share owned by each member. The first item issued by the Combine was *The English Pilot, Mediterranean* since Seller had most of it ready for publication. John Thornton supplied some plates for the volume and others were made in his shop. After this volume was issued, the Combine reissued the *Coasting Pilot,* the *Atlas Maritimus,* and the *Description.* In doing so, they altered the imprints on the engraved title insets and recut the London View title plate to prolong its printing life.

After two years the Combine was dissolved so the common property was distributed. William Fisher obtained the rights to *The English Pilot, Southern* and the *Atlas Maritimus* which he entered in the *Stationer's Register* on November 30, 1679. He

53 W. R. Chaplin, "A Seventeenth-Century Chart Publisher," *American Neptune* 8 (1948), 302–304.

also received several map plates, the London View title plate, some of the engraved title insets, and the printed sheets for *The English Pilot, West India*. James Atkinson took some printed maps, copies of the London View title plate, and the engraved title inset. He altered the imprint on the inset and published his own version of the *Atlas Maritimus*.[54] John Colson appears to have obtained nothing as he probably had the least invested. John Thornton retained his own plates and received some of Seller's. Most probably he also received the printed sheets for *The English Pilot, Oriental*.

In 1689, Thornton and Fisher used the sheets of *The English Pilot, West India* obtained from Seller and completed that volume of the series. In 1703, John Thornton issued *The English Pilot, Oriental* but did not use the sheets Seller had printed earlier, although they did serve as the basis for his own text.[55]

With the formation of the Combine, Seller lost nearly everything he had accumulated, but when it was disbanded he attempted to recover his earlier status. In the *London Gazette* for November 24–27, 1679, he announced his plan to produce an *Atlas Anglicanus* with John Oliver and Richard Palmer.[56] As previously, Seller prepared a new engraved title plate and inset along with several maps before failure again defeated him. As part of this scheme he petitioned the Lords of the Treasury for permission to import 10,500 reams of elephant paper free of duty.[57] After this scheme collapsed, Seller concentrated on publishing smaller items that he could manage. For the most part these were minature editions of his earlier works and were usually made up of maps and other illustrations printed from the same plates. Among such can be listed the *Atlas Maritimus*, the *Atlas Minimus*, the *Atlas Coelestis*, the *Anglia Contracta*, the *Atlas Terrestris*, *A Pocket Book*, *An Almanack for an Age*, *A New System of Geography*, and numerous other items.

In the dissolution of the Combine, Seller retained the remainder stock of the text sheets from *The English Pilot, Northern* and *Southern*, and the *Coasting Pilot* as well as impressions from his Dutch plates. None of this would have been of interest

---

[54] Library of Congress, Phillips 4150, 4151. P. L. Phillips was not aware of the sequence in the publication of the *Atlas Maritimus* and misdated these copies.

[55] R. A. Skelton and Coolie Verner, "Bibliographical Note" in the facsimile of *The English Pilot. The Third Book* . . . (Amsterdam, 1970).

[56] R. A. Skelton, *County Atlases of the British Isles 1579–1703* (London, 1970), pp. 186 ff.

[57] S. P. Dom. Entry Book, 55, p. 43.

to the other members of the group. In later years, he attempted to gather new editions of the *Coasting Pilot* and *The English Pilot, Northern* along with the *Atlas Maritimus*, but this effort also failed.

The "old worn Dutch copper plates" were made for an oblong atlas and measured 25.0 cm by 55.0 cm. In order to conserve paper, Seller decided to print two plates on one folio sheet, so he reduced most of them to 17.0 cm by 53.0 cm.[58] This reduction made the charts appear unfinished, but it did not materially affect the content. Not all of his plates were from the Jansson atlas. Some were small folio plates, and, as was customary, Seller cut out of the plate that portion which he needed.

In adapting the plates, Seller deleted the original title in Dutch and replaced it with his own title and imprint in English. A second title in French on many plates was left untouched. It appears that this substitution of an English for a Dutch title did not occur until after the publication of *The English Pilot, Northern*, 1668, for Pepys remarked on that edition that the plates were "without a Dutch word so much as turned into English. . . ." One chart, *Cust van / VLArNDrREN / . . .*, is found in this original form in a copy of *The English Pilot, Southern*, 1672.[59] Later it became Seller's chart of Flanders[60] which seems to verify Pepys's statement about the 1668 edition. In most cases the original scales were also removed and replaced.

Among the charts printed from the Dutch plates there are a number of anomalies that indicate Seller was printing from the plates as he was reworking them. The chart of the . . . *West part of England* . . . is found with the panel blank that contained the title in French, because the title was deleted in error. Some copies of this chart have one coat of arms in the panel and others have two. The chart of . . . *the Sea coasts of Gallisia* . . . in *The English Pilot, Southern* also had the French title removed by accident. This time the French was replaced although the chart is found in other states, with the French removed, partially replaced, and completely replaced.

Some charts were first printed in their normal size at the top

[58] These sizes are approximate as the plates vary slightly in size and Seller did not reduce all plates to the same scale.

[59] This chart belongs in *The English Pilot, Northern*, but is found misbound in *The English Pilot, Southern* in the Yale University copy of *The English Pilot, Northern* and *Southern*.

[60] A detailed study of these plates is in process.

of a sheet with letterpress text below. These plates were then reduced in size and printed on one sheet with another plate. One plate, Flanders, was used in several different ways: in its original Dutch and with letterpress text as noted above; as an extension plate pasted to the Clarke plate; and then with its own extension plate on a sheet. A similar plate was made by Seller to extend the coverage of the *Draught of the Sands. . . .* Quite often these extensions became detached.

The first Dutch plates used by Seller were adapted before March 24, 1671 and do not have the designation "Hydrographer to the King." These plates were used largely in *The English Pilot, Northern* and the *Coasting Pilot.* After that date, Seller added hydrographer after his name in the imprint on subsequent plates. This is useful in determining the adaptation of the plates, but it is not entirely reliable evidence as some plates may have had the title added later.

After he began to expand and develop *The English Pilot, Northern* with a view to publishing a series covering the "whole world," Seller began to prepare folio plates. The first of these was the Clarke plate, discussed earlier, followed by the Lamb plate. Both of these plates have the notation "Hydrographer to the King" which suggests that they were made after March 1671. Since the Lamb plate was issued with the *Description* dated 1671 and it is presumed that the Clarke plate was issued earlier, the title hydrographer must have been added to the Clarke plate and possibly also to the Lamb plate. The *Description* was advertised in February and does not have hydrographer after Seller's name on the title page, so the earlier issue of the chart with the *Description* may not have had it either. Of course Seller may have advertised the *Description* in February but without actually having it ready for distribution until after March 24, 1671.

Following these two charts Seller prepared additional folio plates (fig. 7). In most cases these were copied from charts issued by Dutch publishers. Koeman notes that Seller used charts by Pieter Goos.[61] In the "Directions for the Book-binder to place the Charts . . ." in *The English Pilot, Northern* 1671, there are three folio charts listed: *Northern Navigation, North Sea,* and the *Baltic Sea.* Of these, the first two have hydrographer but the last does not. It is difficult to ascertain but the Baltic chart was probably a folio Dutch chart adapted before March 24, 1671.

[61] C. Koeman, *Maps and Charts of Surinam Lands with the Past* (Amsterdam, 1973), IV, 199.

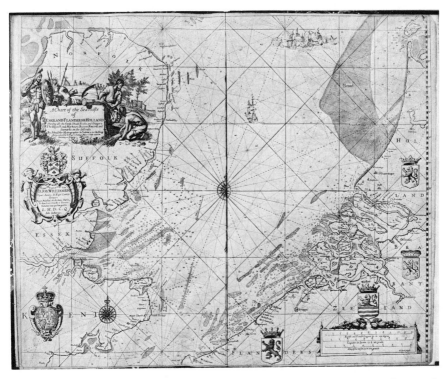

Fig. 7. Folio plate, *A Chart of the Seacoasts of England, Flanders and Holland* . . . c. 1672 by John Seller.

Most copies of *The English Pilot, Northern,* 1671 have additional folio charts not called for in the "Directions. . . ." Among the earlier of these were Tyne / Humber, Russia, and the River Dwina. Other charts were added to copies, gathered later, as they were available. The inclusion of folio charts should provide an indication of the order of gatherings of extant copies of *The English Pilot, Northern* but this is not always reliable because of latter-day book dealer augmentations.

John Seller was inclined to print large runs of his text and plates. Either he overprinted them or he had fewer sales than expected, for he obviously had large remainder stocks. As late as 1701 when Jeremiah Seller and Charles Price were printing the charts for *The English Pilot, Africa,* they used thin paper that they strengthened by pasting impressions to remainder sheets pulled by Seller from his Dutch plates some thirty years earlier. In his later years, Seller himself collected copies of certain items from this remainder stock of sheets.

In gathering copies of his publications from time to time, Seller drew charts from stock without regard for the variance in impressions from a single plate. This practice led to anomalies and inconsistencies among copies that cannot be explained otherwise.

After Seller had made a new plate for Sir Jonas Moore's chart of the Thames Estuary he rewrote his earlier six page *Appendix* that accompanied the Clarke chart and expanded it into ten pages. This was then wholly or partially reset and issued separately as the *Description of the Sands* published in 1671. In both the *Appendix* and the *Description* the text was keyed by letters to the chart. The *Description* was advertised in the *Term Catalogue* for Hilary Term (February 13) 1671. It was also advertised in the *London Gazette,* number 554, March 6 to 9, 1671. The notice in the *Term Catalogue* was as follows:

A Description of the Sands, Shoals, Buoyes, Beacons, Roads, Channels, Sea-Marks, on the Coasts of *England,* from the *South Foreland* to *Orfordness*; shewing their bearings and distances from the most emminent places on the land, with the depth of Water on them as well as in the Channels between them. Being accommodated with a new and exact draught of the Sands, according to the said Description. Published by approbation and license of his Royal Highness the Duke of York. In Folio. Price stitched 3s. Printed for Jo. Seller at the Marriner's Compass in *Wapping* at the Hermitage Stairs.

In the *London Gazette* the advertisement said "There is extant a new description . . ." that seems to indicate the precise date of publication. This issue is known only by a typographic title page, the chart in the first state, and one leaf of text, all of which were found misbound in a copy of *The English Pilot, Northern*, 1671. The imprint indicates that it was printed by John Darby and sold by John Seller in Wapping and by John Wingfield "right against St. *Olaves* Church in *Crutched-Fryars*." This is the same bookseller listed in the imprint in Seller's *Praxis Nautica* published in 1669 which had been entered in the *Stationer's Register* on November 22, 1669 by Wingfield. He is also listed in the imprint of *The English Pilot, Northern*, 1671. It is unlikely that he had any investment in the *Description* as he would also have entered it in the *Stationer's Register*. Wingfield died shortly after *The English Pilot, Northern*, 1671 was published and his widow transferred title to *Praxis Nautica* to William Fisher.[62] After this edition of the *Description* Seller again reset the text into eight pages and added it to *The English Pilot, Northern*, 1671.

The earliest known edition of the *Coasting Pilot* was published in 1672. Seller advertised a *Sea Atlas* and an *English Sea Atlas* that was described as "Discovering the Sea-Coasts in most of the known parts of the World, in General and Particular Charts. . . ." He may have started to produce his *Sea Atlas* but converted it into the *Coasting Pilot* because of the lack of suitable charts.

This volume includes charts and sailing directions for the east and south coasts of England and the coasts of Europe bordering on the Channel. All of the charts were used also in *The English Pilot, Northern*, although some of the Dutch plates occur in different combinations. The charts of Bristol and of Lands End in that combination have been found only in the *Coasting Pilot* as is the case with the charts of Maes and Holland. The Dutch chart of the north coast of England printed on a sheet with text has not been found in the *Coasting Pilot* but is in *The English Pilot, Northern*—suggesting that the remainder stock of this variant form was exhausted by the time Seller was compiling the *Coasting Pilot*. The chart of Weser printed alone on a sheet is found in both the *Coasting Pilot* and *The English Pilot, Northern*.

EPILOGUE

The inability of John Seller to complete his work successfully

---

[62] The entry is in the *Stationer's Register* for 16 July 1673 and 20 November 1675.

can be attributed to a number of factors apart from the obvious lack of adequate capital. He seemed to have no conception of his own limitations, and his ambitions greatly exceeded his abilities. His primary difficulty was his basic dishonesty and lack of integrity in spite of his protestations that he "thought it my duty, for the service of my Countrey, to adventure on this great Trouble and Charge. . . ."[63]

Seller advertised *The English Pilot, Northern* as "Being furnished with New and Exact Draughts, Charts and Descriptions; gathered from the Experience and Practice of divers Able and Expert Navigators of our English Nation,"[64] although he knew he was using old Dutch plates and translating the text from Dutch originals. The plates of 1620 were copies of those made in 1608 and were long obsolete. Even his own folio charts were copies from the Dutch but advertised as new surveys so that he willingly misled mariners who might trust his charts to be new information about waters that were perilous to navigate.

John Seller had little experience in publishing so he constantly altered his publications in such a way that no two copies were identical. However, because of supreme self-confidence, he attempted to produce everything at once which no reasonable publisher would have considered.

In spite of these things, John Seller did initiate the chart trade in London, but it was left to others to make it successful. The six volumes he began in *The English Pilot* were developed and remained in print through all of the eighteenth century in numerous editions. At present, the editions known of each volume are: sixteen of *The English Pilot, Northern*, twenty-three of *Southern*, thirteen of *Mediterranean*, forty-two of *West India*, thirteen of *Oriental*, and nine of *Africa*. Although each volume was produced at first by a different group, all volumes eventually ended up in the hands of Mount and Page. This firm, like Seller, had little interest in the validity or reliability of their publications. Consequently, they reissued them for a century without any significant alterations or improvement so that much of Seller's original material continued in use in spite of its obsolescence.

From its beginning with Seller, the chart trade in England eventually reached a position of world domination not unlike

---

[63] From the Dedication to James, Duke of York, in *The English Pilot, Northern* (1671).

[64] From the title page of *The English Pilot, Northern* (1671).

that enjoyed by the Dutch at the time when Seller sought to break their monopoly. As the years passed, the accuracy of chart making was developed beyond any expectations of Seller or his contemporaries. Although John Seller may be considered to have initiated this, he scarcely merits the status accorded him in maritime history.

V

ENGLISH CARTOGRAPHY, 1650-1750: A SUMMARY

David A. Woodward

INTRODUCTION

This essay is intended to summarize the growth of mapmaking in England from the middle of the seventeenth century to the middle of the eighteenth century. It specifically excludes marine cartography of the period, which is covered in Professor Verner's essay in this volume. The study does not attempt to make a substantive or bibliographical contribution, but draws together some of the published literature in an overview of the period, and offers some suggestions for future lines of inquiry. The organization of the essay is based on a number of functional divisions, for example, route, military, and county maps, large-scale plans, and small-scale maps of continental and world areas. The treatment of these themes is preceded by a technical introduction, summarizing the main developments in the instrumentation and practice of surveying and map printing and publishing.

The spirit of the Restoration was well reflected in the development of cartography. The new age it heralded was brimming with vitality, enthusiasm, and energy to tackle enormous projects. The supremacy of the Netherlands had been broken on the sea and in America, beginning with the first Navigation Act of 1651, which specified that no goods could be imported into England except in English vessels. This independence from the Dutch was mirrored in the atlas business: the London mapsellers began to replace the works of Joan Blaeu and Jan Jansson with those of British cartographers. With such economic expansion, the demand for practical devices, including maps, expanded. In particular, the increase in road traffic required more accurate route maps. The growing interest in the natural sciences led to a desire to map the earth in a more systematic fashion than had been attempted in previous centuries. The eighteenth century became a period of intense activity in cartography as nations became aware of the importance of national systematic topographic map series based on first order geodetic triangulation. The basis of such work lay in a more accurate and precise knowledge of the size and shape of the earth. Sir Isaac Newton's postulate that the earth is an oblate spheroid was upheld by the two French expeditions of Maupertuis and La Condamine that provided estimates for the length of a degree in various latitudes which were to form the basis of eighteenth-century cartography.

TECHNICAL INTRODUCTION

Most studies in the history of cartography have concerned themselves with either the mapmaker or the map itself. There has been very little explanation of the methods by which maps were made or how the techniques have affected the content of maps. Yet, it is clear that the surveying, engraving, and publishing aspects of the history of cartography are integral and vital parts of our knowledge if we are to understand the whole development of a given period of mapmaking.

Surveying. Since cartographers generally use the term surveying as synonymous with *land surveying*, it is important to realize that the latter is only a small part of the former. Thus, in this first text on surveying printed in English, Anthony Fitzherbert's *Here Begynneth a right Frutefull Mater: and hath to name the boke Surveyenge and Improvements* (London, 1523), the author does not give any instructions for actual measurement, and most of the book deals with the value of crops and farm management, or, in short, estate management. In the period covered by this essay, increases in accuracy were largely due to improved practice rather than the development of better instruments. By 1616 the surveyor had the following major tools: the theodolite, described by Leonard Digges in 1571, capable of angular measurement in both the horizontal and vertical plane;[1] the circumferentor, which was used for horizontal sighting only; the plane table; and surveyors' poles and cords. Aaron Rathborne's *The Surveyor*

[1] Leonard Digges, *A Geometrical Treatise Named Pantometria* (London, 1571).

in Foure Bookes (London, 1616) not only summarized the practice of the day, but introduced a number of new elements (fig. 1). The importance of Rathborne's book lies in his clear account of the instruments used and the suggestion that the decimal chain replace the surveyor's rods, anticipating a development by Edmund Gunter in 1620. The unit was one statute perch of 198 inches, a tenth of which was called a prime of 19.8 inches, and a hundredth was called a second, of 1.98 inches. As an aside, it is not generally known that the origin of the symbols for degrees, minutes, and seconds—° (zero), ′ (one), and ″ (two)—lies in the numbered hierarchy of units as used by Rathborne. According to A. W. Richeson, Rathborne's book was probably not used so much by practicing surveyors as by later writers on surveying, but through the latter it had a considerable influence through the seventeenth and eighteenth centuries.[2] Thus William Leybourn's *The Compleat Surveyor* (1653) presented Rathborne in more simple language and probably exerted a greater influence than the model itself. Perhaps the most popular of the surveying manuals of the seventeenth century was John Love's *Geodaesia* (1688), which went into eleven editions between 1688 and 1792. It explained the surveying procedures very clearly and no doubt helped to raise the level of surveying practice of the time.

Even by the end of the seventeenth century, angle-measuring instruments such as the theodolite lacked precision. Telescopes were poor and the vernier scale, although invented in 1631, was not commonly used. The spirit level, apparently invented by Melchisedech Thévenot in 1666, was adopted very slowly, and a method of precise division of scales had yet to be developed. Jonathan Sisson's theodolite of the 1720s, which may have utilized the achromatic lens invented by Chester Moor Hall, was therefore a great improvement. The telescope was shorter, and the bubble level as well as the pinion and vernier scale were employed. It was a more satisfactory and compact instrument than had been available hitherto, but the theodolite was still not a particularly reliable instrument. It was not until toward the end of the eighteenth century that use of large precise theodolites, based on the development of the dividing engine by Jesse Ramsden, was initiated.

[2] A. W. Richeson, *English Land Measuring to 1800: Instruments and Practice* (Cambridge, Mass., 1966), p. 113. Dr. Peter Eden, University of Leicester, is compiling a "Dictionary of Land Surveyors of Great Britain and Ireland (1550–1850)" which, when published, promises to be an invaluable aid to studies of this topic.

Fig. 1. Title page from A. Rathborne, *The Surveyor in Foure Books* (London, 1616). Courtesy of the Newberry Library, Chicago.

Very little research has been undertaken on the effects of these improvements on cartography during this period. There are at least two approaches to this problem. First, one could carry out a Francis Parkman or Thor Heyerdahl procedure and reconstruct the activity using the tools and practices of the period. If an area could be found in which the surveyor and the exact kind of instrumentation were known, a researcher could go into the field with such instruments, and resurvey the piece of land. Obviously, this would have to be in an area that had not changed substantially. The findings could then be compared with the original survey to give an approximate idea of the care and knowledge necessary to obtain a particular result. A second method involves a statistical comparison of the accuracy of maps of a given area, developed experimentally, using different techniques such as that of Fritz Bönisch involving the calculation of mean square errors.[3] Other possibilities include the use of a distortion grid,[4] or vector analysis which can provide an excellent picture of changes in accuracy over a map, and from map to map.[5] Until more objective studies are carried out it is difficult to generalize on the accuracy of the surveying of the period.

Engraving. While the woodcut technique continued throughout this period for specialized use in books, periodicals, and newspapers, engraving on copper was the dominant method of reproducing maps. The latter technique, which has been well summarized by Coolie Verner in *Five Centuries of Map Printing*,[6] had remained unchanged since its introduction in the fifteenth century. This involved three basic problems: (1) to transfer the drawing on the copper, (2) to engrave the plate so that it printed clearly, and (3) to enable corrections to be made with a minimum of effort. Our basic knowledge of the practice in England during this time is largely drawn from William Faithorne's *The*

[3] Fritz Bönisch, "The geometrical accuracy of 16th and 17th century topographical surveys," *Imago Mundi* 21 (1967), 62–69.

[4] Examples of this method are found in Leo Weisz, *Die Schweiz auf alten Karten* (Zurich, 1945), p. 217; Edward Imhof, "Beitrage zur Geschichte der topographischen Kartographie," *International Yearbook of Cartography* 4 (1964), 129–152; J. H. Andrews, "Charles Vallancey and the Map of Ireland," *Geographical Journal* 132 (1966), 55; W. R. Tobler, "Medieval Distortions: The Projections of Ancient Maps," Association of American Geographers, *Annals* 56 (1966), 351–360.

[5] For example, see William Ravenhill and Andrew Gilg, "The Accuracy of Early Maps? Towards a Computer Aided Method," *Cartographic Journal* 11 (1974), 48–52.

[6] Coolie Verner, "Copperplate printing," in David Woodward, ed., *Five Centuries of Map Printing* (Chicago, 1975), pp. 51–75.

Art of Graveing and Etching (1662). The book was inspired by Abraham Bosse's *Traicté de Manières de Graver* (Paris, 1645).

The method of transferring the drawing to the copper plate depended on whether a reduction was necessary, or if the original was expendable. Since photography was not available at the time, the reduction was generally achieved by gridding or with the use of a pantograph, the latter having been first described in 1631. Apparently this instrument was in common use in the early eighteenth century, as seen from its inclusion in Chambers's *Cyclopedia* (1727–1741).

If reduction and preservation of the original were not considerations, the drawing could be varnished to make the design visible from the back. The front of the drawing could then be covered with chalk, placed face down on the copper plate, and traced through with a sharp point such as an etching needle. The copper plate itself had to be coated with a thin layer of wax in order to receive the image. The same result could be achieved with tracing paper or chalk-covered paper placed between the original and the plate. The scarcity of actual drawings used to make engravings may suggest that the originals were routinely destroyed in the process.

The basic tool of the copper engraver was the burin, or graver. This is a small steel rod four to five inches long with a square or lozenge-shaped end set into a wooden handle, the round part of which is fitted against the palm. As the tool is pushed along the surface of the copper through the layer of wax bearing the design, it produces a burr or a thin rough ridge along the edge of the engraved line. This is removed with a scraper. Thin lines were made with the tint tool with a fine triangular cross section, while thicker lines were produced with a scooper (scauper). There was considerable versatility in the thickness of lines controlled by the engraver. The burin could be leaned slightly to one side to produce a thicker line, and as such a leaning was natural in the engraving of curves, a slight thickening of line is often observable in such cases. Tones were produced either with fine hatching using a tint tool, or with a dotted pattern by means of flicks made with a curved graver, or dotting with a hand punch or a punch and hammer. Symbols and lettering were usually hand engraved rather than punched, and the lettering was carefully built up with a variety of tools as described by Diderot.[7]

[7] Denis Diderot, *Encyclopédie, ou Dictionnaire Raisonné des Sciences* (Paris, 1751).

It is likely that dividers were used to scratch parallel guidelines for the lettering of the plate.

Corrections were straightforward in the copperplate process. If it were a small area, the part to be changed was scraped down or burnished flat, while the whole areas could be rubbed down with an oil-rubber (a roll of woollen cloth bound with string) and emery or other polishing powder. Especially thick lines were closed on the plate with a rounded head cocking punch. After smoothing the detail over in this way, the corresponding part of the back of the copper plate was located with the use of long calipers and the back of the plate was beaten up with a hammer or a flat punch to level the printing surface of the copper. The plate was then ready for reengraving.

After engraving, impressions were pulled in a rolling press. Taking an impression from a copper plate is an extremely painstaking process that requires thorough inking by hand using a leather dabber to insure that the ink is forced into all the engraved lines. This procedure, which is done while the plate is warm in order to loosen the ink, has to be repeated after each impression.

The usual copperplate press or rolling press works according to the mangle principle in contrast to the common printing press of the time which used vertical pressure. The paper is dampened, placed against the plate, and both are passed through the press once. Under great pressure, the paper is forced into the hollows, pulling out the ink, producing a brilliant black impression that is characteristic of the method. The plate mark normally seen around the maps is a witness to the amount of pressure required. As can be imagined, the repeated pressure to which the plate is subjected results in the rapid deterioration of the quality of the impression. After about 3,000 impressions the plate usually needs to be deepened by reengraving or replaced by a new plate.[8]

The difficulty of transferring the design, the actual cost of the materials, and the time taken in engraving all contributed to the heavy capital expense of producing an original copper engraving.

[8] Estimates of the number of impressions possible from a copper plate vary widely according to the depth of engraving, hardness of the copper, and so on. F. de Dainville, in *Le Langage des Géographes* (Paris, 1964), p. 74 mentions a figure of 10,000 good impressions. Vittorio de Zonca, *Nuovo Teatro di Machine* . . . (Padova, 1607) estimates 1,000 impressions. The Ordnance Survey, in *Account of the Methods and Processes* . . ., 2d ed. (London, 1902) indicated that a copper plate with fine work would yield about 500 good impressions. R. A. Skelton, in *County Atlases of the British Isles 1579–1703* (London, 1970), p. 36, gives a figure of 2,000 to 3,000 impressions.

In 1679, a piece of copper for John Adams's map of England and Wales cost £1 5s. Gregory King, the engraver of the map, charged Adams £26 8s. for both sheets which took him about three months to engrave.[9] For engraving a small county map, £8 was a normal figure, while the total production cost of the average map might be £20. Since they were sold at sixpence uncolored, a shilling colored, the publisher would have to sell about 400 colored impressions to recover his manufacturing costs. The result of this large capital expense was twofold. First, publishers were dissuaded from engraving original maps but encouraged to buy old copper plates and to print restrikes from them; the oft quoted example of the maps of Christopher Saxton, engraved in the 1570s and reprinted until around 1770, serves to illustrate this fact.[10] Secondly, while there was essentially no legal protection in the way of copyright against piracy of parts of a printed map, the ownership of the plate, representing in itself such a large capital investment, guaranteed a more effective copyright protection than did registration of the document at Stationer's Hall.[11]

Coloring. The existence of several manuals on hand coloring of prints and maps bears witness that this craft was extremely important in this period.[12] The technique used was *limning* which was essentially a water color wash. The manuals contained recipes for mixing various colors, and suggestions concerning which colors should be used for particular features. For example, William Salmon recommends "green walnuts" as an excellent color to express highways, lanes, etc. *The Complete Academy* contains instructions on how to prepare the maps for the color wash:

[9] E. G. R. Taylor, "Notes on John Adams and Contemporary Map Makers," *Geographical Journal* 97 (1941), 182–194. For other figures relating to engraving costs, see Sarah Tyacke, "Map-sellers and the London map trade c. 1650–1710," in Helen Wallis and Sarah Tyacke, eds., *My Head is a Map: A Festschrift for R. V. Tooley* (London, 1973), pp. 10–74.

[10] Reprinted by C. Dicey & Co. See H. Whitaker, "The later editions of Saxton's maps," *Imago Mundi* 3 (1939), 72–86.

[11] R. A. Skelton, *County Atlases*, p. 231.

[12] For example, *Albert Dürer revived: or a book of drawing, limning. washing or colouring of maps or prints* (London, 1666); *The Complete Academy: or a Drawing Book, containing . . . the plainest method for colouring maps and prints*, 2d ed. (London, 1672); William Salmon, *Polygraphice* (London, 1672); John Smith, *The Art of Painting in Oyl*, 3d impression (London, 1701); Henry Wilson, *Surveying improv'd, with directions for making transparent colours for maps* (London, 1725).

When your paper is pasted upon the Cloath, and dry, then wash it over with white starch made very weak, and that will both receive your Colours and keep them when they are layd. For the using of Allum-water it is ridiculous, for it will not only be a means to cockle the paper, and bring all out of proportion, but to be sure it will starve the Colours in a short time, notwithstanding it does promise so much at first.[13]

As in other aspects of map production at the beginning of this period, England looked to the Netherlands to provide a technical model for map coloring. Thus, in John Smith's manual *The Art of Painting in Oyl*, we are told that "the only way to colour Maps well is by a pattern done by some good Workman, of which the *Dutch* are esteemed the best."[14] By the middle of the seventeenth century, the art of coloring on Dutch maps had reached its height, and the naturalistic ornament around the map groups (figures, ribbons, flowers, etc.) were prime subjects for the colorist's art. Near the end of this time, however, a definite trend toward the use of color as a method of conveying information rather than as a decorative aid may be seen. Intricate engraving and ingenious systems of symbolization were no longer obscured by thick washes, and color was frequently reserved only for boundaries and coastlines. Many maps, however, remained uncolored and color printing of maps was virtually unknown.

The Map Trade. If there is one theme that captures the spirit of the map trade in the century from 1650 to 1750, it is that of growing independence from the mapsellers, engravers, and papermakers of the Continent. From the middle of the seventeenth century to the beginning of the eighteenth, there was a great expansion in the market for prints, and especially maps, growing from a social, intellectual, and economic demand. Such a demand had to be met by an increase in the number of engravers and printsellers as well as in the domestic supply of raw materials for papermaking.

In the sixteenth and seventeenth centuries most paper used for maps was imported. The last quarter of the seventeenth century marked a turning point in the pattern of supply. In 1650 there were thirty-seven paper mills in England; by 1670 there were probably fifty, clustered mainly in the Home Counties to avail themselves of the center of rag supplies and the market of London which, between 1699 and 1703, accounted for 97 percent of

---

[13] *Complete Academy,* p. 42.
[14] John Smith, *The Art of Painting in Oyl,* p. 106.

imported white paper.[15] By 1720, England was producing about two-thirds of her needs, thus helping to stabilize the supply.

The translation of Abraham Bosse's manual of engraving by William Faithorne in 1672 was but one indication of the newly found independence of the English engravers.[16] While the demand for prints was met to some extent in the last quarter of the seventeenth century by Dutch sheet maps of Visscher, Danckaerts, and deWit, and while Dutch plates were still used (as in Moses Pitt's *English Atlas*), more English maps were engraved ab initio using new information. The independence of English cartography in this period was expressed in a number of highly imaginative and important enterprises such as John Ogilby's road maps, the several projects for an English atlas, the interest in triangulation schemes for the mapping of Britain such as that propounded by John Adams, new county surveys, and an increase in the number of general world atlases.

Another change was seen in the pattern of the print trade toward the end of the seventeenth century. Printsellers became increasingly specialized and independent of booksellers. Whereas the printer and the bookseller were rarely the same person, map engravers of the time, because of the high cost of engraving and the capital required for stocking sheets for assembly into atlases, often published and sold their wares. The great majority of maps published between 1660 and 1700 came from the shops of no more than a dozen publishers: William Fisher, Christopher Browne, William Berry, John Garrett, Philip Lea, John Ogilby and William Morgan, Robert Morden, Richard Mount, John Seller, John Thornton, and Robert Walton. Many of these mapsellers consolidated to publish specific works by taking shares in jointly owned plates. Most of the printsellers were clustered in the City, but after the Great Fire in 1666, a westward trend could be seen as mapsellers moved into the Holborn, Fleet Street, and Charing Cross districts.[17]

ROUTE MAPS

Route maps indicated how to get from A to B, and included side features only when they helped to identify the route. In this respect, they differed from general purpose maps intended to

[15] D. C. Coleman, *The British Paper Industry 1495–1860* (Oxford, 1958), pp. 3–23.
[16] William Faithorne, *The Art of Graveing and Etching* (London, 1662).
[17] R. A. Skelton, *County Atlases*, p. 236.

illustrate the spatial distribution of a variety of geographic features. The road map, or more specifically here the "strip map," has had a great deal of treatment in the literature, especially in the work of Sir H. G. Fordham and more recently in that of J. B. Harley. The strip map concept relates to the Roman road map known as the Peutinger Table.[18]

Roads appeared on British maps as early as 1360 (the Gough map in the Bodleian Library, Oxford).[19] But it was not until the end of the sixteenth century that roads appeared on printed maps such as those by John Norden of Middlesex (1593) and Hertfordshire (1598) and that of Kent by Philip Symonson (1596). The general county atlases and maps in use at the time, such as those of Christopher Saxton and John Speed, and the Dutch county maps of Jan Jansson and Joan Blaeu, showed no roads.

The publication of *Britannia Volume I* by John Ogilby in 1675 was a master stroke of concept and execution (fig. 2). It was the first major advance since the work of Christopher Saxton and is one of the landmarks in British cartography. By arranging the work according to a strip map format and permitting no extraneous information, Ogilby succeeded in covering much of the most populated part of England and Wales at the medium scale of 1 inch to 1 mile. Unlike his predecessors, he used the standard mile of 1,760 yards throughout the work. As Harley indicates, Ogilby's *Britannia* well reflects the spirit of the Restoration.[20] In particular, the ambitiousness of the age is clearly seen in the atlas, for the *Britannia* was only part of a much larger projected work, namely a world atlas and geography. The *Britannia*, which was intended to cover 40,000 miles of roads, only covered 7,519. While this in itself was a spectacular achievement, it illustrates how far the work fell short of the grand plan.

John Ogilby was born in Edinburgh in 1600 and was, among other things, a dancing teacher and performer. He turned to publishing later in life and, like many others, lost his entire stock in the Great Fire of London in 1666. He was sixty-nine when his major work was proposed. In the same year he and William Morgan were in charge of surveying the new boundaries of the fire-damaged city and in 1676 he produced a superior map of

[18] R. A. Skelton, *Maps; A Historical Survey of Their Study and Collecting* (Chicago, 1972), pp. 41–42.

[19] E. J. S. Parsons, *The Map of Great Britain circa A.D. 1360 Known as the Gough Map* ... (Oxford, 1958).

[20] J. B. Harley, "Bibliographical Note," in *John Ogilby: Britannia, London, 1675* (Amsterdam, 1970).

Fig. 2. Frontispiece from John Ogilby, *Britannia: Volume the First* (London, 1675). Courtesy the Newberry Library, Chicago.

London. These early cartographic efforts introduced Ogilby to a number of people who contributed to the compilation of his *Britannia*—Robert Hooke, curator of experiments of the Royal Society; and the surveyors William Leybourn, John Holwell, and Gregory King. Unfortunately most of Hooke's support went to Robert Plot who, like Ogilby, had proposed a general survey of the entire country.

Ogilby's maps are all on a scale of 1 inch to 1 mile, and he uses the English mile of 5,280 feet where 16½ feet = 1 pole, 40 poles = 1 furlong and 8 furlongs = 1 mile. Miles are symbolized on the maps by two small dots and furlongs by a single dot (fig. 3). Ogilby's scheme of surveying England and Wales from the roads proved to be a very practical method on a modest budget. It is clear that he was aware of the importance of his survey in improving the general survey of England, for he hoped that his work would contribute to a more accurate measurement of the length of a degree of latitude.

Ogilby's methods of surveying are not set out precisely in his introduction to the *Britannia*, but it is clear from the several illustrations of measuring wheels that he used an instrument (the "waywiser") invented by Robert Hooke. It is probable that, in conjunction with this instrument, Ogilby's surveyors used the magnetic compass, chain, and quadrant to perform what was in fact a compass traverse along each road, taking in the details on either side. While only two of the maps are actually signed, namely by Joseph Moxon and Wenceslaus Hollar, the variations in engraving style point to other engravers. Indeed Gregory King claims to have engraved three or four of the maps, and the vigorous style of Moxon's engraving is seen in many of the plates.

Three editions of the *Britannia* appeared in 1675, each with minor changes. In 1676 the work was issued in separates at sixpence a sheet, and an edition appeared with a shortened version of the London text. A year later Swale and Morden brought out a new edition with the text reset. Harley has demonstrated how Ogilby's *Britannia* can be used as historical evidence.[21] Detailed regional maps of the period are lacking and even the evidence on Ogilby's maps should be used in conjunction with available written sources. Nevertheless, much useful information can be derived firsthand from the maps. Evidence can be found on road conditions, early tollgates, details on land use, the extent of built-

21 Ibid., pp. xix–xxiii.

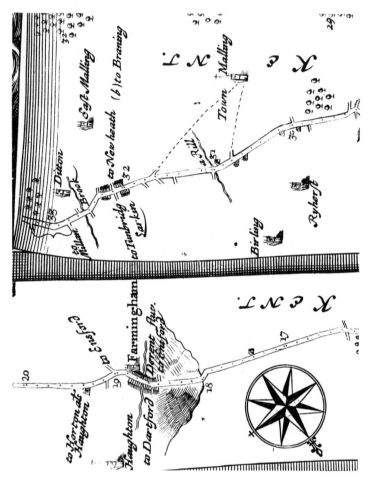

Fig. 3. Detail from J. Ogilby, Britannia, Volume The First (London, 1675).
Courtesy of the Newberry Library, Chicago.

up areas, and villages that have since disappeared. The road maps provide an indication of the position of the road before diversions in turnpiking and before the enclosure of common land. Through small naturalistic symbols, information on the construction material of bridges and the location of windmills and water mills is provided.

In a broader context, E. C. K. Gonner has calculated the approximate amount of common and enclosed land in each county in the seventeenth century by using Ogilby's distinction between "open" and "closed" roads on the maps.[22] The *Britannia* also provides evidence of the length of the old English mile, which was in use before the statute mile of 1,760 yards came into general use. While the latter was made legal in 1593, at first it was only used in London and the surrounding area. Ogilby's three sets of measurements given in his text, horizontal distance (as the crow flies), old English miles, and statute miles, provide what is perhaps the best guide to these early measures.[23]

Derivatives. While John Ogilby's *Britannia* appeared in only six versions before 1700, its impact on later work was considerable. Many roadbooks in the eighteenth century were derived directly from it, and many more depended in some fashion on the content or concept of Ogilby's work.

One reason for the small number of editions of the *Britannia* may be due to its unwieldy format. It is a folio volume 21 inches high and could not therefore be easily carried while traveling. Thus it is not surprising that the direct extant derivatives from the *Britannia* are without exception in pocket form. The text was issued separately in 1699 as a traveler's guide, and the maps appeared in versions by Thomas Gardner in 1719, John Senex (also in 1719), and Emanuel Bowen (1720). Later in the eighteenth century these works were superseded by those of Daniel Paterson and John Cary. A French edition of the Senex version was published in 1759. Plagiarism of Ogilby's road data began immediately, with Thomas Basset and Richard Chiswell reprinting the road tables in John Speed's *Theatre*. The road information also found its way onto Robert Morden's playing cards and on Philip Lea's and Herman Moll's county maps. Ogilby's concept was applied to other road atlases elsewhere in the eighteenth century such as those of George Taylor and Andrew Skinner in Ireland,

[22] E. C. K. Gonner, *Common Land and Inclosure* (London, 1951).
[23] J. B. Harley, "Bibliographical Note," p. xxii.

Christopher Colles in North America, and George Taylor in Scotland.[24]

Later versions of Ogilby's road maps include the unsuccessful edition by Herman Moll—only two specimen pages are preserved and the scheme was shelved. Thomas Gardner's *A Pocket Guide to the English Traveler* (1719) reveals careless copying and rushed engraving, reflecting Gardner's character of the business-man and opportunist rather than the professional cartographer. That of John Senex, a well-established cartographer, is titled *An Actual Survey of All The Principal Roads* and also appeared in 1719.

The most successful version was that of Emanuel Bowen and John Owen, *Britannia Depicta*, which appeared in 1720.[25] Bowen's project may have been a direct descendant of Moll's work, since a proposal for a similar scheme, probably issued in 1718, links Moll to Bowles, Overton, and Bowen in a publishers' consortium. The *Britannia Depicta*, one of the most popular road atlases of the eighteenth century, was basically a reduced copy of Ogilby's *Britannia* and acts as a steppingstone to the roadbooks of Armstrong, Cary, and Paterson. It was Bowen's first major cartographic assignment as he engraved everything, includ-ing the text. The maps appeared in the same sequence as those in Ogilby's *Britannia* at half the scale (½ inch to 1 mile). Seventeenth-century landowners' names were deleted and new information provided primarily from the landed gentry was added. Although only twenty new place names were added, a major improvement was the formalized written spelling rather than the transcribed phonetic form.

THE ROAD DISTANCE MAPS OF JOHN ADAMS

The diagrammatic distance maps of John Adams are probably based on Ogilby's data. There are two versions of this map. A large twelve-sheet edition has two known states, one dedicated

---

[24] See Sir H. George Fordham, *The Road-Books & Itineraries of Great Britain, 1570–1850* (Cambridge, 1924); R. H. Fairclough, " 'Sketches of the Roads in Scotland, 1785'; The Manuscript Roadbook of George Taylor," unpublished; Walter W. Ristow, ed., *A Survey of the Roads of the United States of America 1789 by Christopher Colles* (Cambridge, Mass., 1961).

[25] J. B. Harley, Introduction to facsimile of Emanuel Bowen, *Britannia Depicta* (London, 1720; Newcastle, 1970).

to Charles II bearing Adams's name, and a later state in 1696 by Philip Lea.[26] In addition, there is a small two-sheet version of this map: *Angliae Totius Tabula* . . . , which first appeared in 1679 (fig. 4) and later in some copies of John Adams's *Index Villaris* (1680).[27]

We do not know whether the distances marked on the Adams maps represent road or direct distances, but we assume the double straight lines between places represent the existence of a road. Nevertheless, this matter needs to be researched further and a closer comparison between the large and small versions of the map needs to be made.

The smaller version had a number of derivatives, including a one-sheet map of England by N. Fisher and J. Overton, a two-sheet map of England and Wales by W. Berry, a two-sheet map by C. Browne, a one-sheet map of the cities of England and Wales by R. Walton and R. Morden, and a one-sheet map of England and Wales by H. Overton. These maps, plus the Philip Lea version of the larger Adams map were recently acquired by the Clark Library and form an impressive series.[28]

In 1681, John Adams approached the Royal Society with a proposal for a triangulation based on precise astronomical measurements, but he failed to obtain any financial commitment from the society. While the project failed through lack of funds, Adams is to be credited with an early concept of a national triangulation system based on astronomical observations.

LARGE-SCALE PLANS

The considerable improvement in the instrumentation and practice of land surveying and in the education of surveyors throughout this period was reflected in an increase in the number of large-scale plans for specific purposes. We do not know precisely how many there were, for there have been very few censuses compiled, but the record offices, the copyright libraries, and the older university libraries in England have them in great num-

[26] Edward Heawood, "John Adams and his Map of England," *Geographical Journal* 79 (1932), 37–44.
[27] E. G. R. Taylor, "Notes on John Adams" (see n. 9), and N. J. W. Thrower, "Seventeenth-Century Distance Maps by John Adams," *UCLA Librarian* 25 (1972), 23–25.
[28] Thrower, "Seventeenth-Century Distance Maps," p. 25.

Fig. 4. Distance map of England, Wales and southern Scotland in two sheets by John Adams, 1679. Courtesy of the William Andrews Clark Memorial Library, UCLA.

bers.[29] They are mainly manuscripts based on simple land survey, done with a chain and plane table, or compass traverse (fig. 5). These "estate plans" were created for purposes such as the sale, maintenance, and development of landed or real property. Other such works included parish or parochial maps, diagrams illustrating proposed enclosures and manorial plans.[30]

There is a surprising lack of literature on the history of the large-scale plan in England. While there are many studies on specific plans and mapmakers, synthetic investigations on the methods and practices used in compiling large-scale maps are hard to find. There appear to be studies on the history of surveying, and considerable research on manorial or estate management in the period, but very little exists on the actual construction methods that were employed.

Town plans form a distinct group of large-scale maps. Darlington and Howgego give one of the few synopses of town mapping in England during the period in their introduction to a fine bibliography of maps of London.[31] They describe the plan of Ogilby and Morgan printed in 1682, commissioned after the Great Fire of London, showing the destroyed areas. No new survey was made until that of John Rocque some sixty years later. In the early years of the eighteenth century, the trade relied on new editions of old maps mainly emanating from the Netherlands. Rocque produced two major maps of London—one, at a scale of 5 inches to 1 mile, of the City, and one of the environs (fig. 6). It was not until the end of the eighteenth century that

[29] Published lists include Elizabeth M. Rodgers, *The Large Scale County Maps of the British Isles 1596–1850*, 2d ed. (Oxford, 1972); F. G. Emmison, *Catalogue of Maps in Essex Record Office, 1566–1855* (Chelmsford, 1947; supplements 1952 and 1964); Ian H. Adams, *Descriptive List of Plans in the Scottish Record Office*, vol. 1 (Edinburgh, 1966), and *The Mapping of a Scottish Estate* (Edinburgh, 1971); Elizabeth M. Elvey, *A Hand-List of Buckinghamshire Estate Maps* (Buckinghamshire Record Society, 1963); *A Union Catalogue of Large Scale Manuscript Maps of Scotland* is in process of compilation, see *Cartographic Journal* 3 (1966), 56. For other works, see J. B. Harley, *Maps for the Local Historian. A Guide to British Sources* (London, 1972).

[30] For definitions, see the correspondence between Alan Gillies and R. A. Skelton, *Cartographic Journal* 4 (1967), 53, 140. A good example of an intensive study of a manorial plan is William Ravenhill, "The Mapping of Great Haseley and Latchford: An Episode in the Surveying Career of Joel Gascoyne," *Cartographic Journal* 10 (1973), 105–111.

[31] Ira Darlington and James Howgego, *Printed Maps of London circa 1553–1850* (London, 1964). For a discussion of the use of early town plans to reconstruct the topography of English towns, see M. D. Lobel, "The value of early maps as evidence for the topography of English towns," *Imago Mundi* 22 (1969), 50–61; and M. D. Lobel, ed., *Historic Towns*, vol. 1 (London, 1969).

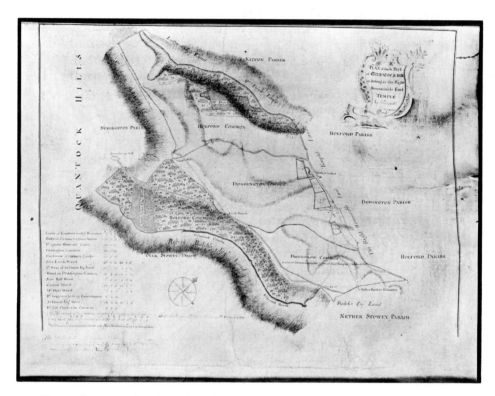

Fig. 5. Large-scale plan of Dodington, Quantock Hills, Somerset belonging to Earl Temple. This plan, dated 1765, and signed, W. Fairchild, Surveyor, is a compass traverse with precise area measurements and magnetic variation (declination) allowance. Courtesy of the Huntington Library, San Marino, California.

Fig. 6. Detail of a map of the environs of London by John Rocque
c. 1742. Courtesy of William Andrews Clark Memorial Library,
UCLA.

the largest plan of the century, that of Richard Horwood, was
to appear.

MILITARY MAPS

Unlike the estate surveyor or the cartographer of town plans, the
military surveyor was interested not so much in property as in
terrain. While the location of houses, inns, and public buildings
might be of value from the point of view of quartering troops,
property boundaries could be easily crossed and were commonly
disregarded. But the representation of relief is obviously an important characteristic of this class of map.

While this period ends before the initiation of the Ordnance
Survey of England and Wales at the end of the eighteenth century, there were enterprises that may be precursors of the Ordnance Survey. One such enterprise, described by Skelton, was
the military survey of Scotland.[32] Its significance lies in its completeness in giving a historical crosssection of the area. Following
the rebellion that broke out in the Highlands in 1745 and cul-
minating in the battle of Culloden in the following year, it became clear to the British government that, owing to the inac-
cessibility and remoteness of the area, a map would be of great
value. While the work is not technically exceptional from either
a surveying or cartographic point of view, it is typical of military
mapping in the mid-eighteenth century. The survey was initiated
in 1742 by Lt. Colonel David Watson who was its director until
1754. During this time, William Roy, who was to lead the initial
efforts of the Ordnance Survey later in the eighteenth century,
was an assistant quartermaster engaged in the field work.

The instruments employed were circumferentors with non-
telescopic sights, and chains 45 or 50 feet in length. The relief
features and boundaries of settlements and rural properties were
sketched in by eye or apparently copied from existing maps. But
as can be seen from fig. 7, the relief was executed in a fairly
elaborate fashion, following the style of the Cassini surveys in
France—a great improvement over the symbolic sketches of
earlier printed maps in England. Roy described this work as
"a magnificent military sketch rather than a very accurate map
of a country." He ascribed the deficiencies to the use of inferior
instruments and to the lack of funds. But while the map was "a

[32] R. A. Skelton, *The Military Survey of Scotland 1747–1755*, Royal Scottish
Geographical Society, Publication no. 1 (Edinburgh, 1967).

Fig. 7. Detail of the Crieff area from the MS survey of Scotland. From R. A. Skelton, "The Military Survey of Scotland 1747–1755," Royal Scottish Geographical Society *Special Publication No. 1.* Courtesy of the Society and the Trustees of the British Museum.

military sketch," it provides a fairly realistic impression of the surface of the terrain which makes up in part for the lack of planimetric accuracy.

MULTI-SHEET COUNTY MAPS

The term "multi-sheet county map" is intended here to mean separately published maps of English counties on scales of about 1 or 2 inches to a mile. They exclude the county maps found in the so-called county atlases of England and Wales described by Chubb and later by Skelton, which are collections of maps of English counties.[33]

The maps under consideration are, for instance, those described in the listing of Elizabeth Rodgers.[34] From 1676, following John Seller, most of the maps were based on the meridian of St. Paul's Cathedral in London, and from 1721, following John Senex, the maps employed the English mile of 69½ miles to a degree. The first maps that fall under this category were the one-inch maps made in the early eighteenth century, such as those by Gascoyne (Cornwall, 1700), Williams (Denbigh and Flint, c. 1720), Budgen (Sussex, 1724), Beighton (Warwickshire, 1721—the first of a number to be based on triangulation), Senex (Surrey, 1729), and that of Christopher Packe (East Kent, 1736), which also enjoys the distinction of being the first map to show spot heights as a means of terrain representation.[35]

An impetus to the improvement in county maps in the middle of the eighteenth century was provided by the premiums offered by the Society of Arts, as described by Harley.[36] The quotation from a 1755 letter of one of the initiators of the premiums, William Borlase, sheds light on the state of county maps at the end of this period:

[33] It should be pointed out that the American use of the term "county atlas" is quite different from the English. The American county atlases were atlases of specific counties, not collections of maps of various counties. See Norman J. W. Thrower, "The County Atlas of the United States," *Surveying and Mapping* 21 (1961), 365–373.

[34] Elizabeth M. Rodgers, *The Large Scale County Maps*, see n. 29.

[35] Eila M. J. Campbell, "An English Philosophico-Chorographical Chart," *Imago Mundi* 6 (1949), 79–84.

[36] J. B. Harley, "The Society of Arts and the Surveys of English Counties 1759–1809," Royal Society of Arts, *Journal* (1963), 43–46; (1964), 119–123, and 269–275. See also J. B. Harley, "The re-mapping of England, 1750–1800," *Imago Mundi* 19 (1965), 56–67.

I would submit to you as a friend, whether the state of British Geography be not very low, and at present wholly destitute of any public encouragement. Maps of England and its counties are extremely defective. We have but one good county map that I know, and the headlands towards all of our shores are at this time disputed and even where Halley himself made his observations.[37]

A similar, more succinct view was offered by a correspondent to the *Gentleman's Magazine* who referred to "Moll's little erroneous trifles"[38] in 1748 when complaining about the lack of correct maps in northwest England.

The premiums were announced in 1759 and were to grant £100 for any "approved" county map on the scale of 1 inch to a mile or larger. For a profile or section of a navigable river, there was to be an additional gratuity. The better class of maps falling into this category were produced after the middle of the eighteenth century. These include the many maps by Thomas Jefferys (e.g., Yorkshire, 1771–1772), a map of Berkshire by John Rocque (1770), the Andrews, Dury, and Herbert map of Kent (1769), and many others (fig. 8).

SMALL-SCALE MAPS

The group of maps under consideration in this section constitute those of continental and world areas. It is at these small scales that the problems of projection and generalization become more apparent. The smaller the scale, the more there is to select and simplify, and the larger the area to be covered by the map, the more obvious the transformation from sphere to plane.

During the period covered by this essay there were very few advances in map projection. The Classical projections, such as the stereographic, gnomonic, and the three projections of Ptolemy, were already in use, as were the sixteenth-century cordiform projections. Mercator's projection (1569) was, of course, popular for marine charts, but was also commonly used for general world maps. The sinusoidal projection, attributed to Nicholas Sanson (c. 1650) proved to be a valuable equal-area projection and was frequently employed for large areas near the equator.

We may recognize several groups of small-scale maps that

---

[37] Harley, "Society of Arts," p. 43.
[38] *Gentleman's Magazine* 18 (1748), 1.

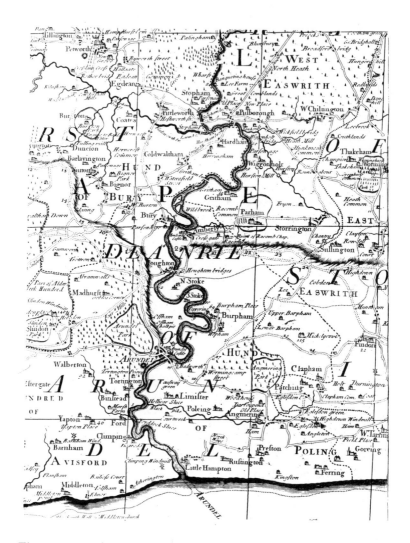

Fig. 8. Detail from R. Budgen (Map of Sussex), 1724. Courtesy the Bodleian Library, Oxford.

were made during this period. Some are in general atlases and are exemplified by the work of Herman Moll, John Senex, and Moses Pitt. There are maps in geography books and accompanying atlases as well as the small maps in magazines designed to keep the reader up-to-date with current affairs. Lastly, although they do not form a large group, there are the imaginary maps such as those found in some novels of the period.

Contemporary sources in which such maps are described include the sale catalogs of the period, such as those by Robert Scott who issued four between 1664 and 1687. Moses Pitt, whose atlas we shall discuss later, produced a catalog in 1674. Other catalogs include those by Benjamin Walford, who had been an apprentice to Robert Scott, Samuel Smith, who had been Moses Pitt's apprentice, and Samuel Buckley, who issued annual catalogs from 1695–1700 and subsequently became editor of the *London Gazette*, 1714–1742.[39] Twenty-nine map-sellers' catalogs from the period 1655–1720 are known.[40]

The *Term Catalogues* are another extremely useful contemporary source for listing maps. They appeared regularly from 1668 through 1709 and are summarized in the four cumulative lists printed by Robert Clavell in 1673, 1675, 1680, and 1696, reprinted and edited by Edward Arber in 1903–1906.[41] The *Term Catalogues* are the first source to provide publishers' names and prices.

Contemporary broadsides from map publishers are few and apparently only three are known: those of Peter Stent, 1649–1653 and 1662, and that of John Overton, published in 1674.

GENERAL ATLASES

During this period there were many unfinished projects for atlases covering the whole of Britain and in some cases attempting to cover the entire world. Most of these were failures. Richard Blome's unsuccessful project was derived largely from Camden and Speed. John Ogilby's atlas, which had started so

[39] Graham Pollard and Albert Ehrman, *The Distribution of Books by Catalogue* (Cambridge, 1965).

[40] For further information on this subject, see Sarah Tyacke, "Map-Sellers and the London Map Trade c. 1650–1710," in Wallis and Tyacke, eds., *My Head is a Map*, pp. 63–64. A photocopy collection of mapsellers' catalogs of this period belonging to the late R. A. Skelton is in the Map Library of the British Library.

[41] Edward Arber, *The Term Catalogues, 1668–1709 . . . ,* 3 vols. (London, 1903–1906).

well with the *Britannia*, comprised only two other volumes, one each of Africa and America which, because of lack of data, were not up to the high standard of the original volume. John Seller's *Atlas Anglicanus* was also a commercial failure. Perhaps the most lavish disappointment was the atlas of Moses Pitt, whose proposal of 1678, based on the concept patterned on the *Grand Atlas* of Joan Blaeu, called for eleven or twelve volumes (fig. 9).[42] Four volumes appeared, with two printed from worn Jansson plates originally bought by Johan Waesberg and Stephen Swart. Swart was a correspondent of Pitt's and saw a chance of selling the plates. Largely through the contact of Robert Hooke, Pitt was able to gain the approval of the Royal Society for the project. It was extremely ambitious, with 600 maps on 900 sheets, with full text descriptions—a major geography. Pitt was committed by his prospectus to deliver the work by March 24, 1679, at the price of 40s a volume. At that time, he asked for a year's grace, after which he still failed to produce. Volume 1 appeared in 1680, volume 2 in 1681, volume 4 in 1682, and volume 3 in 1683. His main difficulty was in finding people to write the descriptions. Here, the Oxford school of geography came to his aid. Obadiah Walker, Hugh Todd, William Nicholson, and Richard Peers all contributed descriptions of various European countries. Thomas Lane and Robert Hooke helped to edit the work. The only map designed, drawn, and engraved in England was a map of the polar regions, perhaps by Michael Burghers. Only one new British county map, which appears in an early eighteenth-century Dutch "atlas factice," was produced (that of Angus), engraved with the joint imprint of Waesberg, Swart, and Pitt.

The best general atlas of the period was *The World Described* (fig. 10) by Herman Moll who was born in the Netherlands and emigrated to London around 1681. In that year he engraved maps for Jonas Moore in his *New Systeme of the Mathematicke*.[43] Very little else is known about his life. Moll had a very distinctive style and apparently did most or all of the engraving for his atlases himself.

[42] E. G. R. Taylor, "The English Atlas of Moses Pitt, 1680–83," *Geographical Journal* 95 (1940), 292–299.

[43] J. N. L. Baker, "The Earliest Maps of H. Moll," *Imago Mundi* 2 (1937), 16; Henry Stevens and Henry Robert Peter Stevens, *The World Described . . .* (1952). Typescripts in British Museum and Lilly Library, Indiana University.

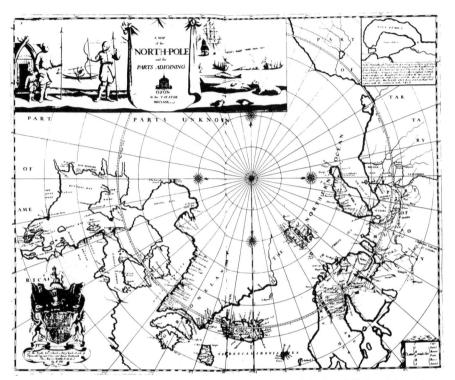

Fig. 9. Map of the North Polar regions presumably by Michael Burghers from the *English Atlas* of Moses Pitt, 1680. Courtesy of Special Collections, UCLA Library.

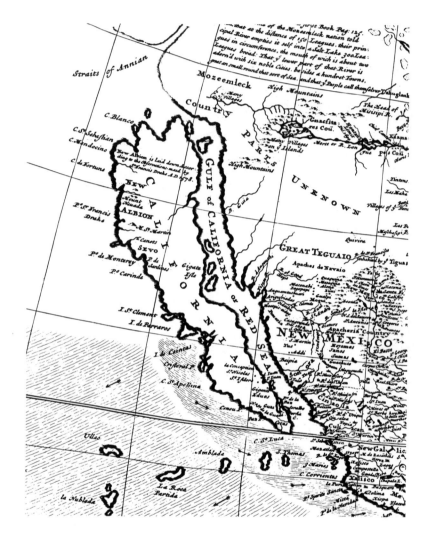

Fig. 10. Detail from Herman Moll, "Map of North America . . ." in *The World Described* (London, c. 1720). Courtesy the Newberry Library, Chicago.

MAPS IN GEOGRAPHIES

During this period a large number of geographical treatises on the use of maps and globes appeared. They involved the disciplines of astronomy, geometry, and geography and were frequently written "for the instruction of ladies." These geographies were often illustrated by small diagrams in the text and folding maps. One such example is that by Thomas Salmon, *A New Geographical and Historical Grammar*, 1749, with plates by the extremely versatile copper engraver, Thomas Jefferys.[44] Salmon apparently "caused a set of new maps to be engraved, that may agree with the work, and corrected them with my own hands; for since the days of my friend Herman Moll, the geographer, we have nothing but copies of foreign maps, by Engravers unskilled in geography, who have copied them with all their errors."[45] Another class of maps in geographies compared classical and modern geography, a model that was soon to become very popular throughout the eighteenth and nineteenth centuries. The best known of these works was by Edward Wells and was reprinted throughout the century.

MAPS IN EIGHTEENTH-CENTURY MAGAZINES

A bibliography, shortly to be published, draws attention to the many maps appearing in English periodicals in the eighteenth century.[46] These magazines were compendia of moral essays, poetry, epigrams, notices, promotions, and, of course, news, both foreign and domestic. Maps were soon added as a selling point. Perhaps the best known of these magazines was the *Gentleman's Magazine*, with its first map appearing in 1736. In 1745, Thomas Jefferys made the acquaintance of Edward Cave, the proprietor, and produced over twelve maps for the magazine between 1746 and 1748. The *London Magazine* followed with maps to maintain the competition, as did the *Universal Magazine of Knowledge and Pleasure*, in 1747, introduced by an enterprising Lon-

---

[44] J. B. Harley, "The bankruptcy of Thomas Jefferys: an episode in the economic history of eighteenth century map-making," *Imago Mundi* 20 (1966), 27–48.

[45] Edward Wells, *A New sett of maps, both of ancient and present geography* . . . (Oxford, 1700).

[46] Christopher M. Klein, "A Checklist of Maps in Eighteenth-Century British Periodicals," forthcoming.

don publisher, John Hinton. Hinton was convinced that the *Gentleman's Magazine* and the *London Magazine* were too well established to have a competitor in the same field and therefore tried a different approach—that of covering technology and general information instead of current events and political affairs. Some time before 1747 he came in contact with Thomas Kitchin and made plans for a county atlas in parts, with a specific program of publication. In the June issue of 1747, the magazine included a map of Berkshire by Kitchin to illustrate an article. Somehow the editors of the *London Magazine* lured Kitchin's atlas away from Hinton, and in November, 1747 a map of Bedfordshire was published in that journal. Edward Cave was outraged. He said, "Nothing is easier than to copy foreign maps, and old descriptions of counties, in most of which are numberless errors, especially with regard to the (county) fairs in so much that many countrymen have rode forty or fifty miles to no purpose."

IMAGINARY MAPS

The appearance of imaginary maps in novels reflects an interesting facet of the character of authors and the reading public at the time. Considering the many eighteenth-century editions of such works as Jonathan Swift's *Gulliver's Travels* (1726) and Daniel Defoe's *Robinson Crusoe* (1719), the exposure of these maps must have been considerable.

As examples of this type, we may discuss the maps in *Gulliver's Travels*, which have been described fully by Bracher and others. The maps in *Gulliver* are a hybrid of the known and imaginary: for example, the peninsula of Brobdingnag is attached to the authentic northwest coast of America (fig. 11). It appears that the authentic parts of the maps were copied from parts of a map by Herman Moll which Bracher identifies as "A New & Correct Map of the Whole World . . ." in *The World Described* [1709–1736?].[47] The cartographer and / or engraver have not been identified, but Bracher believes that the maps were probably originally done without Swift's knowledge, for the geography is inconsistent with the text. But the author must not have disapproved entirely, for, while having every opportunity to do so, he did not have them removed from subsequent editions. Perhaps

[47] Frederick Bracher, "The Maps in *Gulliver's Travels*," *Huntington Library Quarterly* 8 (1945–1946), 59–74.

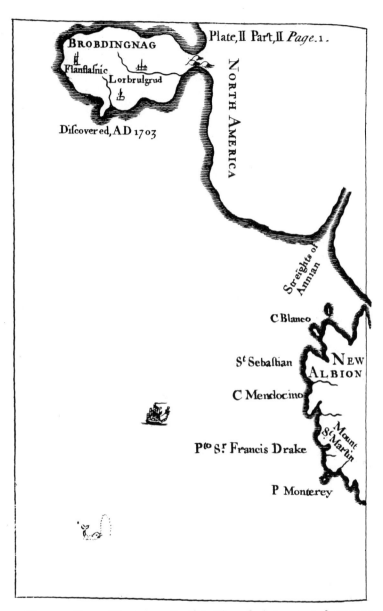

Fig. 11. From [Jonathan Swift], *Travels into several remote nations . . . by Lemuel Gulliver* (London, 1726). Courtesy of the Newberry Library, Chicago.

the maps successfully expressed Swift's disdain for maps. As a
true champion of the liberal arts, he resented what he regarded as
an increase in scientific knowledge (as shown on maps) without
regard to moral improvement of the time. In such a vein, Swift
ironically alludes to Herman Moll in *Gulliver* as "my worthy
friend."

SUMMARY

With the exception of a few highlights such as the work of
Ogilby and Halley, the century 1650–1750 was not a major
period in English cartography. John Ogilby's *Britannia* stands
out as one of the boldest and most imaginative projects of the
period. There were few advances in surveying theory and practice: large-scale plans were surveyed essentially according to the
methods described by Rathborne in the first half of the seventeenth century, while engraving processes remained essentially
unchanged.

The important maps of Edmond Halley, covered in Professor
Thrower's essay, represent highlights in graphic representation
during the period, but others had to await the considerable advances in thematic cartography in the first half of the nineteenth
century. The general reliability and accuracy of the geographical
content of maps saw no great improvement either; compared to
the French efforts of Sanson, Delisle, and the Cassini family, the
English contribution was poor indeed. The Royal Society was
not to have the electrifying effect on cartography in England
that was seen in France under the support of the Académie
Royale. The period ends before the exciting advances made by
British cartographers of the late eighteenth century which were
stimulated at home by the need for large- and medium-scale
multi-sheet maps of counties and abroad by the military needs of
the French and Indian War, the American Revolution, and the
administration of the British Empire. The domestic need was to
result in the establishment of the Ordnance Survey in 1791, while
the overseas efforts were to call upon the talents of such men as
J. F. W. Des Barres, James Rennell, and Captains James Cook
and George Vancouver.

The general mediocrity of the period's cartography may be a
reason for the lack of studies on almost every aspect of this
activity, and while the road mapping of Ogilby is covered relatively fully, there is very little on the actual methods employed

in producing the thousands of large-scale manuscript plans of the period. We have very little biographical information regarding most cartographers of the time. For example, almost nothing is known about the life of Herman Moll, so that it is not possible to evaluate his work except by painstakingly comparing each section of each map with other contemporary sources and drawing specific conclusions. There is a great need in the history of cartography for some systematic method of comparing the representation of a given region on a map with other maps so that its source may be traced. So often, an important map is derived from a number of other maps, which are themselves acquired from other maps. Only by tracing back to these sources can the information be properly evaluated. This might seem to be a basic theme in the history of cartography, but there are very few studies that really attack the problem. The study of what might be called *cartogenealogy* needs to be systematically developed, perhaps along the lines of textual analysis for literary work as described by Professor Dearing.[48] This is an open field, and our knowledge of the forces at work in English cartography would be much stronger if studies could be undertaken along more systematic lines.

[48] Vinton A. Dearing, *A Manual of Textual Analysis* (Berkeley and Los Angeles, 1959). The technique has been adapted to maps by Alexander B. Taylor, in "Name studies in sixteenth century Scottish maps," *Imago Mundi* 19 (1965), 81–89; and J. H. Andrews, "An Elizabethan surveyor and his cartographic progeny," (Trinity College, Dublin).

VI

EDMOND HALLEY AND THEMATIC GEO-CARTOGRAPHY[1]

Norman J. W. Thrower

Maps form an important part of the record of man's achievement, but they are admittedly difficult to preserve, store, and catalogue. Even when bound into atlases they are frequently oversized and require special handling. But the difficulties of dealing with maps have been magnificently overcome by our librarians, and we now have excellent holdings of cartographic materials at the University of California, Los Angeles. I have been delighted at the support given by UCLA to the acquisition of maps and atlases, including recently a complete set of Prince Youssouf Kamal's superb *Monumenta Cartographica Africae et Aegypti*. In the UCLA library are important collections of contemporary atlases in the Reference Department, maps and atlases published since 1900 in the large Map Library, and cartographic materials, generally earlier than this date, in the Department of Special Collections. The Special Collections materials include a fine copy of Ortelius's Atlas of 1579, a set of the *English Atlas* of Moses Pitt (1680), and a map by Edmond Halley, among many other items. The Halley map is from the splendid Shearman Collection of

[1] The author wishes to express his appreciation for a John Simon Guggenheim Memorial Fellowship that enabled him to carry out the research in 1964 on which this paper is largely based. He also wishes to thank the officials of various institutions who rendered valuable assistance, especially those of the Royal Society, the Royal Geographic Society, the British Museum, and the Public Record Office, in London, the Bodleian Library, Oxford, and the National Maritime Museum, Greenwich. See also "Edmond Halley as a Thematic Geo-cartographer," *Annals of the Association of American Geographers*, 59 (1969), 652–676.

early maps and atlases which was purchased with funds generously provided by the Friends of the UCLA Library.[2]

In the cartographic record of the seventeenth and eighteenth
centuries we can recognize diversity through the development of
thematic or special subject maps. This trend, which has continued
at an accelerated rate in recent times, is exemplified by the cartographic contributions of the English natural philosopher Edmond
(Edmund) Halley, Hally, Hailey, or Hawley, as he is variously
known.[3] A thematic map is one designed to illustrate some particular subject or to serve some special purpose; it stands in contrast to a general map on which a variety of phenomena might be
displayed to satisfy a number of uses. Thus among contemporary
sheet maps a topographic map would be a general map, and a soils
map a thematic one. In modern atlases thematic maps are generally found near the front and show such distributions as population, rainfall, etc., and are usually followed by general maps that
often combine on one plate, landforms, lines of transportation,
settlements, political boundaries, etc. Naturally, sometimes the

[2] From time to time, information on cartographic acquisitons of the Uni-
verity of California, Los Angeles, appears in the *UCLA Librarian*, to which I
have contributed the following invited articles: "Monumenta Cartographica
Africae et Aegypti," supplement to 16, 15 (1963), 121–126; "The Shearman
Collection: An Acquisition of Rare Geographical Books and Maps," supplement to 19, 6 (1966), 61–66; and "The English Atlas of Moses Pitt," 20, 1 (1967),
1–3. An important map contained in the Shearman Collection is the subject of
a joint article (with Dr. Young-Il Kim), "Dong-Kook-Yu-Ji-Do: A Recently
Discovered Manuscript Map of Korea," *Imago Mundi* 21 (1967), 30–49. Recent
Clark Library cartographic acquisitions are described in two articles in the
UCLA Librarian: "Seventeenth Century Distance Maps by John Adams," 26,
6 (1972), 23–25, and "Defoe and the 'Atlas Maritimus' " (Maximillian E. Novak,
co-author), 26, 6 (1973), 32–33. A hand-colored copy of Halley's Atlantic chart,
without dedication, and a mid-eighteenth century Dutch version of Halley's
world magnetic chart (not yet the subject of an article) have also been added
recently to the Clark Library collections.

[3] Edmond Halley is the spelling he used when he signed his full name, as on
his will; see *Notes and Queries* 155 (1928), 24–25. Frequently, "Edmund" has
been used by others for his Christian name, and the surname has been rendered
in various ways. For example, his contemporary and associate in the Royal
Society, Robert Hooke, referred to him in his diary as Hally, which is an
earlier form of the family name; see *Notes and Queries* ser. 11, 9 (1914), 106.
Edmond Halley is the manner in which his name was entered in the parish
register of St. James's, Dukes Place, Without Aldgate, where he was married
in April 1682; *Notes and Queries* ser. 11, 4 (1911), 85, 198. Sir William Laird
Clowes, in his *The Royal Navy: A History* (London, 1898), p. 544n, states:
"His name appears in the MS Navy List of 1698 as Edmund Hawley—a form
that indicates its pronunciation." It has also sometimes been spelled Hayley,
Haley, Haly, and even Hawled.

distinction is not altogether sharp, but on thematic maps base data such as coastlines, boundaries, and places are used only as a means of reference for the phenomenon being mapped, not for their own sake. Of course, thematic maps were constructed before Halley's time; for example, land use maps were drawn in England in the Tudor period. Also, the first edition of Ortelius's *Theatrum Orbis Terrarum* from Plantin's press, dated 1579, contains a supplement with maps showing facets of the geography of the past, such as the eastern half of the Mediterranean at the time of St. Paul's missionary journeys. But these examples are of a very different character from the thematic maps of Halley, who illustrated a number of his scientific theories by cartographical means.

English cartographers of the seventeenth century were heirs of two major traditions in map making. There was, in the first place, the legacy of English surveyors of Elizabethan times, including William Borough and Christopher Saxton, and, secondly, the strong influence of hydrographers, cartographers, and map publishers of the Low Countries.[4] The Dutch influence was transmitted in a number of ways, such as through the English translation of the *Mariner's Mirrour* (1588) and of other important geographical compilations and also by the acquisition of Dutch plates for such a work as Moses Pitt's *English Atlas*. The Dutch influence was also felt through Netherlanders working in England and through Englishmen who had returned home from Holland. Michael Burghers, a Dutch engraver who resided in Oxford for much of his life and is usually credited with contributing to the *English Atlas* of Moses Pitt, is an example of the first type. The second may be represented by Joseph Moxon, a repatriated Englishman who received his professional training in the Netherlands. As Carey Bliss pointed out in his Clark Library seminar on Moxon and seventeenth-century printing, "Joseph Moxon was honored for his work as a hydrographer and globe and map maker by being elected to membership in the Royal Society. Joining him in membership in the Society on that day [November 30, 1678] was his good friend Edmund Halley of

---

[4] Borough and Saxton were leaders in marine and land surveying, respectively, in Elizabethan England. Concerning the former, see A. H. W. Robinson, *Marine Cartography in Britain* (Leicester, 1962), pp. 1–33, and for the latter, Sir H. George Fordham, *Some Notable Surveyors and Map-Makers of the Sixteenth, Seventeenth and Eighteenth Centuries and Their Work* (Cambridge, 1929), pp. 1–9.

comet fame."[5] It was through the efforts of a small company of workers, of which Halley is an outstanding member, that a cartography expressing the new scientific spirit was initiated. In addition, through such men, leadership in the field of cartography passed from the Low Countries to Britain and France.

Edmond Halley had a long and eventful life, in the course of which he came into contact with many of his most famous contemporaries. Here it will be possible only to sketch his biography with broad strokes and emphasize the events that relate to his mapping activities. Scholars who have seriously studied Halley have usually expressed surprise that no adequate biography exists of this man who was long regarded as "the second [after Newton] most illustrious of the Anglo-Saxon philosophers."[6] Recently two popular book-length studies have appeared of Halley's life: one by Angus Armitage of London University and the other by Colin Ronan of the Royal Astronomical Society.[7] In the preface of his study, Armitage states that this is a work on Halley which "does not claim to be a definitive biography" but is a historical evaluation "more elaborate than any previously undertaken."[8] Students of Halley in this century, including Ronan and Armitage, have been indebted to the American scholar, Eugene Fairfield MacPike, who collected and published a great deal of material on the scientist's life and work.[9] Included in MacPike's publications is an unsigned memoir on Halley by one of his younger contemporaries; this is thought to be Martin Folkes, president of the Royal Society from 1741 to 1752. The memoir was transcribed by Professor Stephen Rigaud of Oxford, who collected material for a Halley biography that he did not live to complete.[10] It is from such sources, and the

[5] Carey S. Bliss, *Some Aspects of Seventeenth Century Printing with Special Reference to Joseph Moxon* (William Andrews Clark Memorial Library, University of California, Los Angeles, 1965), p. 17.

[6] *Nature* 21 (1880), 303–304.

[7] Angus Armitage, *Edmond Halley* (London, 1966), and Colin A. Ronan, *Edmond Halley: Genius in Eclipse* (London, 1969).

[8] Armitage, *Edmond Halley*, p. xi.

[9] Eugene F. MacPike, *Correspondence and Papers of Edmond Halley* (Oxford, 1932), *Hevelius, Flamsteed and Halley: Three Contemporary Astronomers and their Mutual Relations* (London, 1937) and *Dr. Edmond Halley (1656–1742): A Bibliographical Guide to His Life and Work Arranged Chronologically* (London, 1939). In this last item are listed articles and notes in various journals dealing with the family life and work of Halley, including several by MacPike himself.

[10] Professor Rigaud's son, Rev. Stephen J. Rigaud, who wrote on the history of science, intended to continue this work toward the biography of Halley. It was never completed but the Rigaud papers were deposited in the Bodleian Library, Oxford.

good account in *Biographia Britannica*, that much general information on Halley's life and work is available.[11]

I wish here to concentrate on Halley's maps rather than on the theories on which they are based, because on one hand most of the theories, though ingenious, have been entirely or largely superseded; on the other hand the cartographic innovations have enduring value and are in use today. My paper, then, is not on Halley as an astronomer, geophysicist, or physical geographer, but is rather upon his contributions to thematic geo-cartography.[12] I use the term "geo-cartography" because cartography is not necessarily focused on the earth. There is, of course, a cartography of extraterrestrial phenomena which by no means is a new field. But it is now becoming increasingly important so that it may be desirable to distinguish between terrestrial and extraterrestrial mapping as we distinguish between astronomy and geography. Some of Halley's earliest and most important contributions to astronomy were in charting or mapping the constellations, and it is as an astronomer that we most often think of him. His great work in that field and his connection with Newton are well known and will receive little attention here where our concern is with Halley's work in geo-cartography and especially with five terrestrial, thematic maps of his authorship.

Edmond Halley was born on October 29, 1656, in his father's country house in Shoreditch, then to the northeast of London, but now part of the metropolitan area. His father, also named Edmond Halley, was a prosperous soap manufacturer with a house and premises in the city. In due course, young Edmond Halley was sent to St. Paul's School. This London school had been founded about a century and a half earlier by John Colet, dean of St. Paul's Cathedral and a friend of Erasmus and Sir Thomas More. Its purpose was to promote the Christian humanism of the Renaissance, and it could number among its old boys,

[11] *Biographia Britannica* (London, 1757), IV, 2494–2520. This account appears to be based on information supplied by Halley's son-in-law, Henry Price (d. 1765). A short general article on Halley's scientific work with excellent illustrations is Sir Edward Bullard, "Edmond Halley (1656–1741) [*sic*]," *Endeavour* 15, 60 (October 1956).

[12] Specific aspects of Halley's work have been the subject of separate articles, e.g., Sir Edward Bullard, "Edmond Halley: The First Geophysicist," *Nature* 178 (1956), 891–892, and Sydney Chapman, "Edmond Halley as Physical Geographer and the Story of His Charts," *Occasional Notes of the Royal Astronomical Society*, no. 9 (1941), 15 pp. In his short and informative study, Professor Chapman omits discussion of certain important cartographic works of Halley, as is pointed out in a review of the above paper by Arthur H. Hinks, *Geographical Journal* 98 (1941), 293–296.

John Milton and Samuel Pepys. The High Master of St. Paul's (a title used to this day for the head of the school) during at least Halley's later years there was Dr. Thomas Gale who had resigned the Regius Professorship of Greek at Cambridge to take the post. Later Gale was elected Fellow of the Royal Society, and, as a member of its council, was to play a significant role in his pupil's life.

At St. Paul's, Halley studied classics and mathematics, subjects that included astronomy and navigation. He also constructed scientific instruments and measured the variation of the magnetic compass, a schoolboy observation that, along with others, he later published. Halley became captain of the school and, in the summer of 1673, entered Queen's College, Oxford. He took instruments with him and continued his observations. While still an undergraduate, in 1676, he published the first scientific paper of some eighty he eventually contributed to the *Philosophical Transactions* (later *Philosophical Transactions of the Royal Society*).[13] It was on the problem of determining the orbit of a planet, a geometrical extension of the Keplerian hypothesis.

During the same year that this paper was published, Halley wrote to Henry Oldenburg, the secretary of the then youthful Royal Society, about the feasibility of preparing a star catalogue of the southern hemisphere. It was intended to complement one of the northern hemisphere made possible by the observations of Hevelius in Danzig (Gdansk), Cassini in Paris, and Flamsteed in Greenwich. John Flamsteed had been installed only a short time before as "Astronomical Observator" (first Astronomer Royal), and Halley had worked with Flamsteed. Halley now applied formally to go to St. Helena to carry out his observations, a request made through Sir Jonas Moore and approved by the king, Charles II. A generous allowance from Halley's father paid the young astronomer's expenses. Halley and a friend who served as his assistant were transported in East India Company ships in the course of regular sailings. During a stay of about a year, and under difficult circumstances, Halley determined the positions of over 340 stars. This information was published late in 1678, less than a year after his return from St. Helena, in a catalogue of southern hemisphere constellations, with an accompanying planisphere or star chart.

---

[13] MacPike, *Correspondence*, pp. 272–278, provides a reasonably complete list of Halley's publications in chronological order.

Halley, who appears to have learned much about navigation as well as astronomy as the result of his voyage, had left Oxford before taking a degree. But the work of St. Helena and a presentation copy of the star chart of the southern hemisphere so impressed the king that he recommended that Halley be awarded the master of arts degree. This was conferred on him "without performing any previous or subsequent exercises" on the third of December 1678.[14] Less than a month before this, as was mentioned previously, Halley was elected a Fellow of the Royal Society. During the next year Halley made a trip to Danzig to visit Johannes Hevelius and to discuss with him the relative merits of instruments with telescopic as against open sights.

On his return from the Baltic, Halley spent a year in England at his father's house before setting out on a "Grand Tour" of France and Italy. This occupied him for a year beginning late in 1680. In December of that year he visited Jean Dominique Cassini in Paris, and together they observed a comet and other celestial phenomena at the observatory. Halley lost no opportunity to engage in scientific pursuits on a wide variety of topics on his trip. He seems to have returned from his tour earlier than planned, probably because of his father's growing financial problems. This situation did not prevent him from marrying, in 1682, Mary Tooke, the daughter of an officer of the Exchequer, with whom he lived happily for over fifty years; they had a son, Edmond Jr., who died a year or so before his father, and two daughters who survived him.

Halley continued his astronomical work at his home near London and, later, in London. In 1683 he published the first of two important papers on terrestrial magnetism in the *Philosophical Transactions*,[15] and in the following year he first visited Newton in Cambridge. It was as the result of this epoch-making meeting, and continued encouragement from Halley, that Newton's *Principia* was published three years later. In spite of increasing financial difficulties resulting from his father's death in

14 Bullard, "Edmond Halley," p. 1.

15 Edmond Halley, "A Theory of the Variation of the Magnetical Compass," *Philosophical Transactions*, no. 148 (1683), pp. 208-221. In this paper Halley makes the following statement (p. 220): "After the same manner we might proceed to conclude the variations in other places under or near the Equator: but I purposely leave it for an exercise to the thoughts of the Serious Reader, who is desired to help his imagination, by having a Map or Globe of the Earth; and to mark thereon the Magnetical *Poles* in the *Longitudes* and *Latitudes* I assign them."

1684, Halley paid for the printing of Newton's magnum opus and saw it through the press. Halley undertook this responsibility in spite of the fact that he found it necessary to resign his fellowship in the Royal Society, in 1685, to accept the paid position of clerk (assistant) to the two secretaries of the society, Sir John Hoskins and his old headmaster, Thomas Gale.

During the next year, 1686, Halley published four papers in the *Philosophical Transactions*, one of these being of particular interest to the history of cartography since it is accompanied by what has been called the earliest meteorological chart.[16] The paper, "An Historical Account of the Trade Winds, and Monsoons, observable in the Seas between and near the Tropicks; with an attempt to assign the Phisical cause of the said Winds," was illustrated by a chart (fig. 1), the original of which has an image size of 19¼ by 5¾ inches. It bears no engraver's name and the copies I have seen are very closely cropped except on the side where it is tipped into the binding; because its longer dimension is greater than the page size of the journal, it requires two folds.

In looking through early numbers of the *Philosophical Transactions*, one is impressed by the quality, quantity, and variety of the illustrations; editorial policy in this regard seems to have been more liberal than is true of many scientific publications today. Halley's first important earth map lacks several cartographic elements. It has no expression of scale; actually, the representative fraction is about 1:50,000,000 at the equator. There is no statement of projection, the grid being formed by 15° longtitude lines, which show by Roman numerals hours of earth rotation from London, and 10° latitude lines; this produces a network of rectangles varying in size according to latitude, and these combine to form a cylindrical projection that is not immediately recognizable. Careful analysis of this projection suggests that the Mercator, a favorite with Halley, as we shall see later, was intended.

The map possesses no title but has since been called "Halley's Chart of the Trade Winds." It also has no legend or key to the symbols used, and one must read the text of Halley's article to

16 Chapman, "Edmond Halley as Physical Geographer," p. 3. See also G. Hellman, *Neudrucke von Schriften und Karten uber Meteorologie und Erdmagnetismus*, no. 8 (Berlin, 1897), pp. 1–13 and plates on unnumbered pages. Perhaps this cartographic work of Halley's might more properly be called a climatological chart.

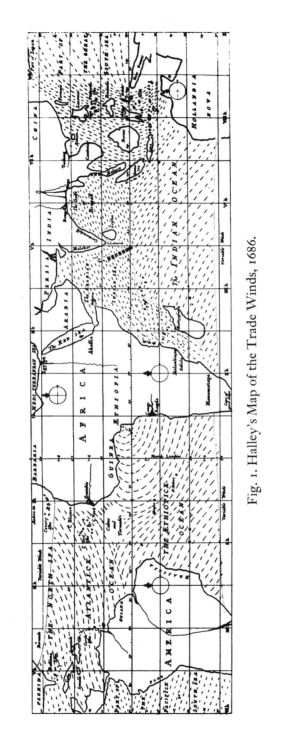

Fig. 1. Halley's Map of the Trade Winds, 1686.

gain this information. The meridian of London (Greenwich) is chosen as the prime meridian; this cartographic convention was rather generally accepted only toward the end of the nineteenth century. The map extends 150° east and 90° west of this meridian and, therefore, covers two-thirds of earth longitude, but excludes most of the Pacific Ocean. It extends through more than 60° of latitude—more than 30° north and 30° south of the equator. One degree ticks (divisions) appear on the equator, the prime meridian, and the marginal meridians.

Coastlines of the area shown reflect the considerable knowledge of the tropics by northern Europeans in the late seventeenth century. A major problem, at least to those not having access to Spanish intelligence, was the geographical relationship of New Guinea and Australia. Halley overcomes this by breaking the coastline between the island and the continent. Except for direction indicators (wind roses), selected place names, and several approximately located rivers, the land areas are left plain.

Most of Halley's paper is concerned with a description of the arrangement of surface winds in various parts of what he calls the Universal Ocean. In particular he discusses the northeast trade winds of the Atlantic north of the equator, and the southeast trade winds of the south Atlantic or Æthiopick Ocean, as he calls it. Recalling his stay in St. Helena, Halley writes, "In this part of the Ocean it has been my good fortune to pass a full year, in an employment that obliged me to regard more than ordinary the Weather."[17] As well as his own experience, he is, of course, dependent on previous descriptions of winds and the observations of sailors. Of this matter Halley says:

if the information I have received be not in all parts Accurate, it has not been for want of inquiry from those I conceived best able to instruct me; and I shall take it for a very great kindness if any Master of a Ship, or other person, well informed of the Nature of the Winds, in any afoermentioned parts of the World, shall please communicate their Observations thereupon; that so what I have collected may be either confirmed or amended, or by the addition of some material Circumstances enlarged. It is not the work of one, nor a few, but a multitude of Observers, to bring together the experience requisite to compose a perfect and complete History of these Winds. . . .[18]

[17] Edmond Halley, "An Historical Account of the Trade Winds, and Monsoons observable in the Seas between and near the Tropicks, with an attempt to assign the Phisical cause of the said Winds," *Philosophical Transactions*, no. 183 (1686), pp. 153–168, esp. p. 155.
[18] Ibid., p. 162.

Paying high tribute to the value of the cartographic method, Halley observed that the "variation [of winds] is better expressed in the Mapp hereto annexed, than it can well be in words."[19] In a later passage he reiterates and underscores this suggestion:

To help the conception of the reader in a matter of so much difficulty, I believe it necessary to adjoyn a Schema, shewing at one view all the various Tracts and Courses of these Winds; whereby 'tis possible the thing may be better understood, than by any verbal description whatsoever.[20]

He then goes on to discuss his method of symbolization quite explicitly:

The limits of these several Tracts, are defined every where by prickt lines, . . . I could think of no better way to design the course of the Winds on the Mapp, than by drawing rows of stroaks in the same line that a Ship would move going alwaies before it; the sharp end of each little stroak pointing out that part of the Horizon, from whence the Wind continually comes; and where there are *Monsoons* the rows of stroaks run alternately backwards and forwards, by which means they are thicker [denser] there than elsewhere.[21]

Explaining why he did not complete his map across the Pacific he adds:

As to the Great South Sea, considering its vast extent, and the little Variety there is in its Winds, and the great *Analogy* between them, and those of the *Atlantic* and *Æthiopick* Oceans, besides that the greatest part thereof is wholly unknown to us; I thought it unnecessary to lengthen the Mapp therewith.[22]

Modern cartographers generally use at least two maps, one each for the extreme seasons of January and July, to show what Halley attempted on one. Accordingly, Halley's map gives too static a delineation of the "perpetual," "general," or "constant" winds, as he called them, or the "prevailing" or "predominant" winds, to use modern terminology. No indication of the movement of the wind belts due to seasonal shifts in insolation is evident on Halley's map. Moreover, it is admittedly difficult to read

[19] Ibid., p. 155.
[20] Ibid., pp. 162–163.
[21] Ibid., p. 163.
[22] Ibid.

direction from Halley's tapering "stroaks"; he employs the now usual and better symbol of arrows only in the area of Cape Verde. Nevertheless, in spite of omitted cartographic elements, its less than perfect symbolization, and its too stationary representation, Halley's map of the winds in the tropics is an original contrbution of great importance to earth science and to cartography.

Halley was well aware of the vital significance of winds to sailing craft from his own experience. It is ironic that adequate instructions for utilizing the winds to the best advantage were available only in the nineteenth century through the work of the American, Matthew Fontaine Maury, at a time when sail was soon to be superseded by steam power, a development that made this information less critical.

Halley's theory of the trade winds presupposes that the heat of the sun in the equatorial regions makes the air there more "rarified" and that air from adjacent regions which is more "ponderous" moves toward it in order to establish equilibrium. Furthermore, the sun's continuous westward movement carries this rarified air along with it at least at the lower levels. Halley suggests that the movement of air from east to west in a vast ocean continues until the return of the sun and so the easterlies are constant.[23] With this thermal explanation of the trade winds, which does not take into account Coriolis Force, Halley joined a number of natural philosophers who had advanced unsuccessful theories to explain the phenomenon. They include Francis Bacon, Galileo, Hooke, and Varenius, whose observations on this subject seem to have been the occasion for Halley's attempt to explain it.

Some years later, in 1699, the English buccaneer, William Dampier, who was a careful observer and reporter of natural phenomena but no great theoretician, wrote his "discourse of Trade-Winds."[24] This descriptive treatise was illustrated with two maps, one of the Atlantic and Indian oceans and one of the Pacific. These maps by Dampier depicting the trade and monsoonal winds were engraved by Herman Moll, a well-known Dutch cartographer who worked in England in the late seventeenth and early eighteenth centuries. In a recent monograph,

[23] Ibid., pp. 165–166.
[24] William Dampier, "Discourse of Trade-Winds, Breezes, Storms, Seasons of the Year, Tides and Currents of the Torrid Zone throughout the World," in *Voyages and Descriptions*. Vol. II, part III (London, 1699), pp. 1–112.

Joseph Shipman correctly challenges the suggestion made by Eva G. R. Taylor that Halley's "pioneer wind map was later used to illustrate William Dampier's useful *Discourse of the Winds* [*sic*] (1699)."[25] However, unlike Shipman, I believe from internal and external evidence that Dampier's trade-wind maps probably owed a good deal to Halley's earlier effort; Dampier refers to Halley in his writings especially in connection with magnetism. Shipman goes on to praise Dampier's maps as "fine examples of draughtsmanship" and to critize Halley's chart as "sketchy and relatively sparse in detail."[26] Actually, Halley's map, with its many discrete symbols each pointing in one direction (even though most of these are not arrows), is conceptually closer to modern wind charts than is Dampier's with its rather vague line shading and scattering of a few arrows. Halley also selected a map projection in which direction can be correctly shown, as on the globe, and where parallels are parallels, allowing easy latitudinal comparisons. These specific properties are not found on the Dampier and Moll maps, on which even the coastlines are not as well drawn as on Halley's map of thirteen years earlier. It is true that Halley did not include information on local winds on his map, but this was not his purpose, nor should we blame him for omitting the winds of the Pacific when he provides excellent reasons for this omission. Halley's suggestions regarding the direction of tropical winds in the Pacific were borne out by the delineation of them by Dampier, who, unlike Halley, had visited the region. It is easy to be attracted to a "pretty" map and overlook the positive qualities of one that is scientifically more valuable but less attractive visually.

Among Halley's professional activities at this period was the editing of the *Philosophical Transactions*. He was intimately concerned with a wide range of scientific information that was then being considered and published, and his own contributions reflect this diversity. Between 1685, when he began his editorial duties, and 1692, when he relinquished them, he wrote on many topics, including eclipses, the motion of projectiles, the time and place of Julius Caesar's invasion of Britain, and the thickness of atoms, as well as upon winds and the variation of the magnetic needle. His second important paper on magnetism appeared in

[25] Joseph Shipman, *William Dampier: Seaman-Scientist*, University of Kansas Publications, Library Series 15 (Lawrence, Kansas, 1962), p. 9.
[26] Ibid.

1692.[27] In the previous year Halley had been considered for the Savilian Professorship of Astronomy at Oxford, his candidacy being supported by a testimonial from the Royal Society prepared by Thomas Gale. But because it was believed that he held unorthodox religious views, Halley was not appointed. Although his editorship of the *Philosophical Transactions* terminated in 1692, Halley's clerkship in the Royal Society carried on beyond this date. His publications continued unabated and include, in this period, studies in the fields of optics, algebra, astronomy, and physics, as well as the construction of mortality tables. In 1696 Halley was appointed deputy controller of the mint at Chester by Newton, who had himself accepted the position of warden of the Royal Mint at the Tower of London. The official duties concerned the standardization and milling of silver coins, but this did not prevent Halley, in the year he was at Chester, from writing articles for the *Philosophical Transactions* on such local topics as the discovery of a Roman altar, an extraordinary hailstorm in Chester, and an eclipse of the moon he observed there. He returned to London about the same time that the Tsar Peter (later known as the Great) took up temporary residence in the country home of the diarist John Evelyn at Deptford. The purpose of Peter's well-known visit was to enable him to learn shipbuilding and other useful arts. Peter sent for Halley and they conversed and dined together; there is even a story that the tsar pushed Halley through Evelyn's holly hedge in a wheelbarrow!

Several years before Halley went to the Chester Mint, negotiations had been in progress at the Royal Society for a ship in which a voyage of scientific discovery and possibly of earth circumnavigation might be undertaken. A small flat-bottomed vessel of 89 tons displacement, 64 feet in length, 18 feet in beam, and 9 feet 7 inches in draught of the type known as a pink, suitable for the purpose, was constructed. A petition was made and, by special order of King William III, Halley was appointed, on August 19, 1698, the commander of the pink, *Paramore* (or *Paramour*) as she was known. He was granted the rank of captain in the Royal Navy, an extraordinary appointment for a man with no previous record in the service. Halley's commission directed him "to call at His Majesty's settlements in America and make some further observations there, in order to better the laying

[27] Edmond Halley, "An Account of the cause of the Change of the Variation of the Magnetical Needle, with an Hypothesis of the Structure of the Internal parts of the Earth," *Philosophical Transactions*, no. 195 (1692), pp. 563–578.

down of longitude and latitude of those places, and to attempt a discovery of what lands lay to the South of the Western ocean"; more important, he was "to seek by observations the discovery of the Rule of the variation of the Compass."[28] By variation was meant, to use Halley's words, "the deflection of the Magnetical needle from the true Meridian [i.e., geographical north]."[29] This is known today as "declination" by surveyors, generally, but still usually as "variation" by sailors.

Since time immemorial, magnetism as a phenomenon has captivated man, and during the later Middle Ages important studies had been made in England on this subject. William Gilbert, who was the climactic figure of Elizabethan science, enunciated the principle that the earth itself is a magnet. However, Gilbert assumed that the magnetic field remained constant or stationary through time.[30] This idea was disproved some years before Halley's birth by Henry Gellibrand, who discovered that "the variation is accompanied with a variation [i.e., declination and secular variation]." Another of Gilbert's assumptions was that the magnetic poles coincide with the geographical poles or poles of rotation. Magnetic observations by explorers in very high latitudes in the seventeenth century indicated the incorrectness of this thesis.

As we have seen, Halley had been interested in magnetism since he was a schoolboy, and two of his most important papers in the *Transactions* concern this subject. In his earlier paper Halley assembled observations from nearly fifty places, some going back almost a century. From these data he attempted to formulate a general magnetic theory and proposed four magnetic poles to explain variation and its change. In his later paper Halley noted that the alteration is a gradual and regular motion. To explain this, he suggested that the earth is composed of an outer shell and an inner globe (or nucleus) which are separated from each other by a fluid medium. The outer shell, he postulated, rotates at a slightly greater rate than the nucleus, and both of these parts of the globe possess two poles each, accounting for the four poles. Halley suggested that, if later observations should prove this explanation to be inadequate, then more internal spheres and more than four poles might be needed. One is re-

[28] MacPike, *Correspondence*, p. 8.
[29] Halley, *Philosophical Transactions*, no. 148, p. 208.
[30] Duane H. Du B. Roller, *The De Magnete of William Gilbert* (Amsterdam, 1959), 196 pp.

minded of the increasingly complicated mathematical models
needed to explain the geocentric (Ptolemaic) astronomical system, when further observations rendered a simple explanation
unsatisfactory. However, Halley offered his ideas on magnetism
only as a "Hypothesis which after Ages may examine, amend or
refute."[31]

The main purpose of Halley's voyage was to test his magnetic theory by providing observations over a large area. A practical consideration was that if an easily measurable relationship
existed between the earth coordinate system and the magnetic
field, more knowledge of this field would help solve the age-old
problem of finding longitude at sea.[32] Halley's voyage has been
described as "the first sea journey undertaken for a purely scientific object."[33] A similar claim is made for Dampier's voyage of
exploration in the *Roebuck* of approximately the same time.[34]
The chronology of these voyages is complicated. Halley left
England in November 1698, two months before Dampier and
returned, having completed his scientific work, half a year before
Dampier. Details of the voyage of the *Paramore* can be read in
the Letters of the Lords of the Admiralty to the Navy Board and
in correspondence between Halley and Josiah Burchett,[35] who
had succeeded Samuel Pepys as secretary of the Navy in 1689.[36]

These letters and the log of the *Paramore*, which was published
long afterwards in Dalrymple's *Voyages*, tell a story of adventure and of devotion to science.[37] To avoid trouble with Moroc-

[31] Halley, *Philosophical Transactions*, no. 195, p. 571.

[32] Norman J. W. Thrower, "The Discovery of the Longtitude: Observations
on Carrying Timepieces at Sea 1530–1770," *Navigation* 5 (1957/58), 375–381,
and "The Art and Science of Navigation in Relation to Geographical Exploration Before 1900," chap. 2 in Herman R. Friis, ed., *The Pacific Basin: A History of Its Geographical Exploration*, American Geographical Society (New
York, 1966), pp. 18–39 and 339–343.

[33] Chapman, "Edmond Halley as Physical Geographer," p. 5.

[34] W. C. D. Dampier-Whetham's review, in *Geographical Journal* 74 (1929),
478–480. In this article by a distant relative of the buccaneer, the *Roebuck*
expedition of 1699 is described as the "first attempt at a voyage planned for the
deliberate purpose of scientific exploration."

[35] This correspondence can be seen at the Public Record Office in London.
It consists of some twenty letters in the "Lords Letter Book," thirty-five in the
"Captains Letter Book," and fifteen in the "Secretaries Letter Book"—all dated
between 16 March 1697/98 and 20 April 1702. There are also "Orders and Instructions" from the Admiralty to the Navy Board etc. Only about one-third
of this correspondence is transcribed in MacPike, *Correspondence*, pp. 103–120.

[36] Clowes, *The Royal Navy*, p. 230.

[37] Alexander Dalrymple, *A Collection of Voyages Chiefly in the Southern
Atlantick Ocean* (London, 1775), pp. 1–22 (Dr. Halley's First Voyage), pp. 1–

can pirates the *Paramore* traveled from England to Madeira in convoy with the squadron of Vice Admiral John Benbow. Benbow, who became a national hero some years later when he continued to fight the French after most of his captains deserted him, returned Halley's initial five-gun salute with five guns as a tribute to science. From Madeira, Halley and his crew sailed southward alone, but when the *Paramore* reached St. Jago in the Cape Verde Islands, she was fired on by two English merchantmen; the masters of these vessels mistook the *Paramore* for a pirate ship. This occurred early in January 1699, when Halley also began to have trouble with his lieutenant. Owing to these circumstances Halley decided to return to England by way of the West Indies earlier than planned. He explains his difficulties with the lieutenant in this way: "because perhaps I have not the whole Sea Dictionary so perfect as he, has for a long time made it his business to represent me to the whole Ships company as a person wholly unqualified for the command their Lopps have given me. . . ."[38] Halley did not realize, until he was at sea, that his lieutenant had proposed an idea for the solution of the longitude problem which had been reviewed and rejected by a committee of which Halley was a member. The resentful officer was courtmartialed, and Halley set sail again in the *Paramore* in September 1699 without a lieutenant. At Portsmouth he called on Admiral Sir Cloudesley Shovell, whose loss of a squadron of ships and two thousand men off the Scilly Isles a few years later was to emphasize dramatically the necessity of finding a means of determining longitude at sea.

On this second voyage, or continuation of the first one, Halley was again escorted to Madeira, and again the *Paramore* continued on alone by way of the Cape Verde Islands southward. The log entry of the *Paramore* for Thursday, December 14, 1699, indicates the careful work being done daily on shipboard:

---

83 (Dr. Halley's Second Voyage), and a map of Trinidada Island. A manuscript version of these voyages, different in many particulars from that used by Dalrymple, is now in the British Museum cataloged as Add. MS 30368, Part I. Part II, not contained in Dalrymple, is the journal of Halley's Third or Channel Voyage. These manuscripts were used by me in my, *The Three Voyages of Edmond Halley in the 'Paramore' 1698–1701* (with a folio of maps), Hakluyt Society, Second Series, forthcoming. Halley almost invariably referred to the vessel as *Paramore* in the official papers, including the title of his three manuscript journals; she was registered as *Paramour* and is also known by this and several other spellings.

38 Dalrymple, *A Collection of Voyages*, p. 22 (Dr. Halley's First Voyage). "Lopps" is Halley's usual abbreviation for "Lordships."

Latitude by a very good observation is 23°.8′. We have had a very fine gale from WSW to WNW and have plied to the Westward all day. True course protracted is 26° S 14 miles, whence difference of Longitude 13 minutes and from London 44°.41′. We are right before the Entrance to Rio [de] Janeiro which at Noon bears from us NNW. Last night the Amplitude was 37°.30′ and this morning 14°.30′.[This refers to the sun's angular distance from the magnetic north measured at sunset and sunrise]. Variation 11°.30′. Let the Longitude of Rio [de] Janeiro be 44°.45′ W from London.[39]

Readings from the barometer and thermometer were also recorded daily, and interesting observations on flora, fauna, and other phenomena were noted as they were seen. By early February 1700 the *Paramore* had reached 52° south latitude where she encountered dense fog and icebergs. Fortunately, the little ship survived these perils, upon which Halley remarks, "God be praised: this danger made my men reflect on the hazards we run . . . amongst these Mountains of Ice in the Fogs which are so thick and frequent here."[40] The *Paramore* now steered northward, passing Tristan de Cunha, St. Helena, and Pernambuco (Recife), where Halley was arrested and held briefly by the self-styled English Consul. After his release they sailed by way of the Antilles to Newfoundland, where they again narrowly escaped shipwreck in the fog and were again taken for pirates and fired on. The *Paramore* finally reached England in early September 1700.

A short time after his return, Halley published, as the result of some 150 observations on magnetic declination taken aboard the *Paramore*, as well as others, a map that is one of the most important in the history of cartography (fig. 2). It is called "A New and Correct Chart shewing the Variations of the Compass in the Western and Southern Oceans" and is credited with being the first printed map showing isogones (i,e., lines of equal magnetic declination). From the writings of Athanasius Kircher we know that an earlier manuscript isogonic chart was drawn, in 1630, by a Jesuit padre of Milan, Christoforo Borri, based on his own observations and those of others.[41] Apparently Borri, in the expectation that the isogones would be useful for navigation, represented them as parallel lines. Kircher was aware that such a

[39] Ibid., p. 22 (Dr. Halley's Second Voyage).
[40] Ibid., p. 35 (Dr. Halley's Second Voyage).
[41] Athanasius Kircher, *Magnes Sive de arte magnetica opus tripartitum,* 2d ed. (Rome, 1643), p. 359 ff.

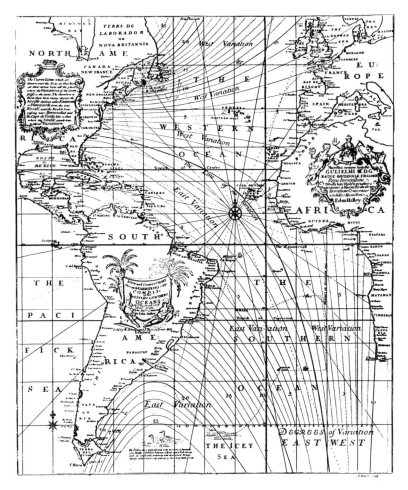

Fig. 2. Halley's Isogonic Map of the Atlantic, 1701. Reproduced by permission of the Royal Geographical Society, London.

simplification was incorrect and seems to have worked on an isogonic map that was never completed, though he hoped to include it with his great study on magnetism. Halley refers to Kircher in his writings but gives no indication that he derived the idea of isogonic lines from this or any other source.[42] Actually, not only is Halley's Atlantic chart the first published map of isogones, which were known as Halleyan (Hallean) lines for about a century, but it appears to be the earliest published isarithmic map of any kind.[43]

In quantitative cartography isarithms (isolines) are lines connecting points of equal intensity of phenomena which have transitional degrees of intensity. In the two and a half centuries since the development of this device there has been an ever increasing number of uses for isarithms in cartography. Some examples include: lines of equal depth (isobaths); lines of equal elevation from a datum, usually mean sea level (contours); lines of equal atmospheric temperature (isotherms); lines of equal travel time (isochrones); lines of equal population density (isodems), etc. Today approximately one hundred different specific isarithms are identified by name, prompting the late Max Eckert, who sired fourteen of these terms, to suggest *iso-nosos*, the disease of inventing terms for isarithmic lines.[44]

Unlike his map of the trade winds, Halley's Atlantic chart was published as a sheet map, not to accompany an article. It measures 22½ by 19 inches and was engraved by John Harris, who was

[42] Halley, *Philosophical Transactions*, no. 148, p. 215. On the description to accompany his Atlantic and World variation charts, Halley states: "What is properly *New*, is the *Curve Lines* drawn over the several Seas to shew the degrees of the Variation of the *Magnetical Needle* or *Sea-Compass;* Which are design'd according to what I my self found in the Western and Southern Oceans in a Voyage I purposely made at the Publick Charge, in the year of our Lord 1700."

[43] Werner Horn, "Die Geschichte der Isarithmenkarten," *Petermanns Geographische Mitteilungen* 103 (1959), 225–232; and S. J. Fockema Andrae and B. van't Hoff, *Geschiedenis Der Kartographie Van Nederland* (s'Gravenhage, 1947), pp. 74–75, 116. Raleigh A. Skelton in correspondence indicated that two manuscript maps with submarine contours (and therefore isarithmic maps) are known to him to antedate Halley's printed map, "A New and Correct Chart Showing the Variations of the Compass in the Western and Southern Oceans" (1700). These are by Pieter Bruinss, 1584, and Pierre Ancelin, 1697. Dr. Skelton added, "It is hard to believe that no cartographer between these two dates thought of joining up points of equal depth; but these are the only ones which have come to light. It is perhaps significant that both these charts are of rivers where the extension of the idea of topographic form-lines to water features might seem natural." The well-known published map of isobaths (submarine contours) of Cruquius was drawn in 1729.

[44] Horn, "Die Geschichte der Isarithmenkarten," p. 226.

employed by Captain Greenvile Collins, the compiler of *Great Britain's Coasting Pilot* (1693). The same publishing house that had handled this important volume, known as William Mount and Thomas Page on Tower Hill at the time of the appearance of Halley's Atlantic chart, sold the map.[45] There is no date, but it is believed to have been published in 1701; the dedication in one of the three cartouches is to King William III, who died on March 8, 1702. In another decorative cartouche is a statement that reads:

The Curve Lines which are drawn over the Seas in this Chart, do shew at one View all the places where the Variation of the Compass is the same; The Numbers to them do shew how many degrees the Needle declines either Eastwards or Westwards from the true North; and the Double Line passing near Burmudas and the Cape de Virde Isles is that where the Needle stands true without Variation [agonic line].

The track of the *Paramore* (on the second Atlantic voyage) is indicated with a dashed line; at its southernmost extent are forms representing icebergs. Nearby are pictures of two sea creatures with the statement:

The Sea in these parts abounds with two sorts of Animalls of a Middle Species between a Bird and a Fish, having necks like Swans and Swimming with their whole Bodyes always under water only putting up their long Necks for Air.

Near the center of the map is an eight-point compass rose with thirty-two rays emanating from it. The only useful purpose this serves on a map of this scale is to indicate cardinal directions, and the rays are distracting. The map extends rather more than 100° W and 20° E of the prime meridian of London with longitudinal gradations along the equator. Latitude is marked on the 20° W line, and the map extends to 59° north and south of the equator. The tropics are indicated by heavy lines. As in the case of the map of the trade winds, the Mercator Chart is used.

To make the map more useful to navigators, an explanation in addition to that provided on the chart itself was needed. A description which was intended to be pasted on the side of the

---

[45] Robinson, *Marine Cartography in Britain*, pp.117–118, gives an account of this firm, including a list of the various names under which the concern did business from 1658 to 1802.

chart was written and signed by Halley. In it he proposes that the projection of the chart, which is commonly called Mercator's, ought to be named "Nautical" because of its special use in navigation. When making this suggestion, Halley may have had in mind that although Mercator, the Fleming, developed this best known of map projections in 1569, it was Edward Wright, the English mathematician, who first published tables for its construction in his *Certaine Errors in Navigation* (1599).[46] Halley notes that the map resulted from the "Voyage I purposely made at the Publick Charge in the year of our Lord 1700." He then cites specific uses of the chart to navigators and indicates the necessity of amending the map because of changes in variation through time. He also asks mariners to inform him of defective observations in the chart.

In the Journal Book of the Royal Society there are a number of scattered references to Halley's varied cartographic activities; the following entry appears under the date October 30, 1700:

Captain Halley produced a Map and Shewed in it his Observations on the variation of the Magnetical Needle which he rectified in the Chart as it was curiously laid down with marks &c in the map.[47]

Since this was recorded only one month after his return to England, we might assume that Halley worked on his map on shipboard and that the map he displayed was in manuscript form. On February 5, 1701, there is another communication on the same subject:

The same [Mr. Halley] shewed a map wherein was the course they had held in several parts of the world; and also the true longitudes and latitudes of many places—he was thanked.[48]

And again, on May 7, 1701:

Mr. Halley tryed the Experiment of the Variation of the Needle this day with the two needles which he had with him in his late

[46] Walter W. R. Ball, *A History of the Study of Mathematics at Cambridge* (Cambridge, 1889), pp. 25–27.
[47] Journal Book of the Royal Society, transcribed from the manuscript copy in the Library of the Royal Society under the dates as indicated. This extract and all other material from the same source is used by permission of the Royal Society, which holds the copyright of the original in the possession of the society.
[48] Ibid.

voyage. And by one the variation was 7°40′ and by the other 8°00′ W.[49]

Yet another entry concerning the map occurs on June 4, 1701:

Mr. Halley presented the Society with a map of his late voyage to the South. He was thanked for it and it was ordered to be hung up in the meeting room.[50]

According to the librarian of the Royal Society, this map, whether in manuscript or printed form, is no longer in the possession of the society. Actually, Halley's Atlantic chart appears to have been lost to cartobibliographers for a long time and was rediscovered in 1895 by the American geophysicist Louis A. Bauer. Bauer reproduced it in the initial issue of *Terrestrial Magnetism*, a journal he founded and first published in 1896.[51] (The copy reproduced as figure 2 is from the Royal Geographical Society; it was acquired for the society as recently as 1920.)

If Halley's Atlantic chart is now rare and was lost sight of for some years, such is not the case with his next important map (fig. 3). This is a world chart showing the same phenomenon as the Atlantic chart, that is, magnetic declination, but extended over a much larger area. Again the engraver was Harris, and the map was sold by Mount and Page. Although at first glance the detail on the Atlantic portion looks the same on both maps, closer inspection indicates that an entirely new engraving was made for the world chart. Nevertheless, some of the remarks we have made concerning the Atlantic chart apply to the world map. The title of the map, in English, contained in a cartouche is "A New and Correct Sea Chart of the Whole World Shewing the Variations of the Compass as they were found in the year MDCC." The same title in Latin appears in another cartouche with the addition of a statement indicating Halley's authorship. In a third cartouche on the land area is the dedication to Prince George of Denmark, Lord High Admiral of England and Generalissimo of

---

[49] Ibid.

[50] Ibid. This probably refers to the engraved and printed Atlantic chart.

[51] Louis A. Bauer, "Halley's Earliest Equal Variation Chart," *Terrestrial Magnetism* 1 (1896), 28–31, and by the same writer, "Some Bibliographical Discoveries in Terrestrial Magnetism," *Nature* 52 (1895), 79–80. Bauer does not seem to have consulted the Royal Society Journal Book, and MacPike gives no entries from this source after February 15, 1689. References 45 to 48 above permit a more exact dating of this map than seems to have been possible previously.

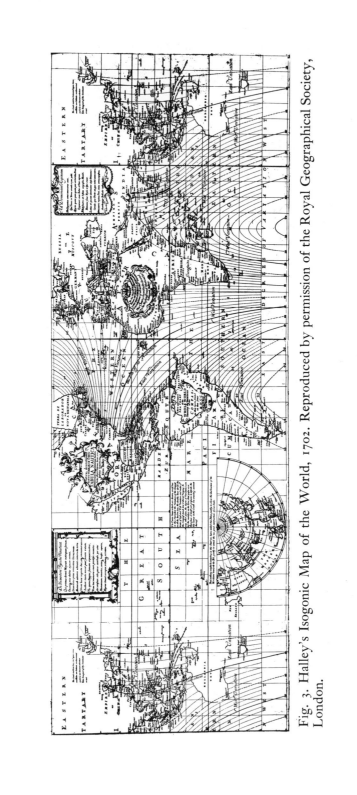

Fig. 3: Halley's Isogonic Map of the World, 1702. Reproduced by permission of the Royal Geographical Society, London.

all Her Majesties Forces; he assumed the latter titles in 1702. This is the year usually assigned as the date of the first edition of Halley's world chart. Prince George was the consort of Queen Anne, who is lauded in a Latin verse, presumably of Halley's composition, contained in a decorative box on the map. Another box frames a Latin verse honoring the unknown inventor of the magnetic compass.[52]

Halley's world chart is a large map measuring some 20 by 57 inches with a scale of 1:33,000,000 at the equator; again the projection is the Mercator with a ten-degree grid. It extends approximately 60° N and 60° S of the equator. Within this latitudinal spread it embraces all of the earth, with the zone from 90° to 160° E being repeated. The most accurate part of the map, the Atlantic area, is in the center, and the lines of magnetic variation have been continued from this region across the Indian Ocean. Concerning this extension, Halley explains in his description below the map that the data have been "Collected from the Comparison of several Journals of Voyages lately made in the *India-Seas*, adapted to the same year [1700]." He excuses himself from continuing the lines across the Pacific with the statement on the map itself, "I durst not presume to describe the like Curves in the South Sea wanting accounts thereof." The line of no variation (agonic line) is plotted both in the Atlantic and through eastern Asia and New Holland (Australia).

Understandably there are inaccuracies in longitudinal position, but most of the coastlines are quite recognizable. Some then recent geographical discoveries are indicated as, for example, "Dampeirs Streights" between New Britain and New Guinea. The more important Torres Strait between New Guinea and Australia, which had been discovered long before by the Spanish, who had suppressed the information, is, of course, not shown. Neither is the south and east coast of Australia and New Zealand, which became known to Europeans through the explorations of Cook and others in the second half of the eighteenth century. Of Hokkaido (Yedso), Halley notes on the map, "it is not known whether Yedso be part of the Continent or not." California is shown as an island, as it was represented for over one hundred years.[53] To illustrate one half of the North Polar regions, he employs part of an azimuthal projection as an inset. The track of

[52] Chapman, "Edmond Halley as Physical Geographer," gives a translation of these verses provided by his wife.

[53] Ronald V. Tooley, *Map Collectors' Circle*, Map Collectors' Series, no. 8, "California as an Island" (London, 1963), pp. 1–28.

the *Paramore* in the Atlantic is not included on the world chart; neither are the wind rose with rays, the sea creatures, or the icebergs. Halley's world chart became well known in Britain and the Continent through its reproduction in a number of atlases; in revised form it continued to be published for over fifty years after his death.[54]

Another continuing theme in Halley's research concerned the tides, which were the subject of one of his earlier papers in the *Philosophical Transactions* (1684).[55] This piece was prompted by a letter on the anomalies of the tides at the port of Batsha in the Gulf of Tonkin, off the south China coast. Halley proposed a solution to the problem related to lunar position, which he illustrates geometrically, to enable navigators to compute the time and height of these tides. But he adds, "to *philosophize* thereon, and to attempt a reason, why the *Moon* should in so particular a manner influence the *waters* in this one place, is a task too hard for my undertaking, especially when I consider how little we have been able to establish a Genuine and satisfactory *Theory* of the *Tides*, found upon our own *Coasts*, of which wee have had so long Experience."[56]

Halley's observations on the tidal phenomenon at Batsha were later discussed in the *Principia* by Newton, who postulated that they could be explained within the existing theoretical framework by the tides passing through different channels at different rates, to produce a complex local situation. When the *Principia* was published in 1687, Halley wrote a letter to accompany the

[54] Halley's World Magnetic map was revised in 1756 and published as "A Correct Chart of the Terraqueous Globe" by William Montaine and James Dodson and continued to be issued until at least 1794. It also appears in later editions of atlases by Seller as well as in those of Ottens, Renard, Philippe de Prétot, et al. Halley edited a reprinted edition of his own and other writings titled *Miscellanea curiosa* (London, Vol. I, 1705, Vol. II, 1706, and Vol. III, 1707). In Volume I opposite page 80, is a world map that combines information from Halley's wind and magnetic charts. It was engraved by Harris and is called "A New and Correct Sea Chart of the Whole World showing the Variations of the Compass as they were found in 1700 with a View of the Generall and Coasting Trade Winds and Monsoons or Shifting Trade Winds by Direction of Capt. Edm. Halley." However, the symbolization of the winds on this map resembles Dampier's chart of this phenomenon more than Halley's earlier effort.

[55] Edmond Halley, "An account of the course of the Tides at Tonqueen in a Letter from Mr. Francis Davenport, July 15, 1678, with a Theory of them, at the Barr of Tonqueen, by the learned Edmund Halley, Fellow of the Royal Society," *Philosophical Transactions*, no. 162 (1684), pp. 671–688.

[56] Ibid., p. 687.

copy presented to King James II.[57] In this letter Halley explains the contents of Newton's opus in simple descriptive terms, especially his theory of the tides. The letter was printed as a tract and reprinted a decade later, in 1697, in the *Philosophical Transactions* without the opening and closing sections.[58] After the general theory is outlined, the tidal problems of specific localities are considered. Halley concludes with a reference to the situation at Batsha, of which he remarks, "that the whole appearance of these strange Tides, is without any forcing naturally deduced from these [Newton's] Principles, and is a great Argument of the certainty of the whole *Theory*."[59]

In view of his long-continued interest in the tides and other phenomena affecting navigation, it is not surprising to find that Halley, a few months after his return from the Atlantic voyage, requested the Lords of the Admiralty to fit out another expedition. He proposed to make "an exact account of the Course of the Tides on or about the Coast of England" for the purpose of improving navigation in these waters.[60] This request met with speedy approval from the Lords, who directed "the *Paramore* at Deptford to be forthwith cleaned and fitted out for sea for Channel service manned with Five and Twenty men and victualled for Three months. . . .[61] Special equipment was provided for the purpose of a coastal survey, but Halley had difficulty finding a suitable crew. In June 1701 the *Paramore* was ready to sail, and in a letter to the Lords of the Admiralty Halley suggested that his instructions should include the following:

> You are to use all possible diligence in observing the Course of the Tides in the Channell of England as well as in mid sea as on both Shores, and to inform your self of the precise times of High and Low Water; of the sett and strength of the Flood and Ebb and how many feet it flows in as many places as may suffice to describe the whole. And where there are irregular and half Tides to be more than

[57] Halley's letter begins, "May It Please the King's Most Excellent Majesty . . ."; it was printed as a broadside in 1687, the year of the presentation.

[58] Edmond Halley, "The true Theory of the Tides, extracted from that admired Treatise of Mr. Isaac Newton, Intituled, Philosophiae Naturalis Principia Mathematica, being a Discourse presented with that Book to the late King James, by Mr. Edmund Halley," *Philosophical Transactions*, no. 226 (1697), pp. 445–457.

[59] Ibid., p. 457.

[60] MacPike, *Correspondence*, p. 116.

[61] Letter from the Lords of the Admiralty to the Navy Board dated 26 April 1701, "Lords Letter Book," Public Record Office, London.

ordinary curious in observing them. You are likewise to take the true barings of the principall head lands on the English Coast one from another, and to continue the Meridian as often as conveniently may be from side to side of the Channell, in order to lay down both coasts truly against one another.[62]

Surveying of coastal areas was no new occupation for Halley. In 1689, a decade before he took command of the *Paramore*, he had produced a chart of the mouth of the River Thames which, according to his own testimony, "corrected several very great, and considerable faults in our Sea-Charts hitherto published."[63] He also made a chart of part of the Sussex coast in 1693. But what he now proposed was a much more ambitious task of surveying English coasts than any he had undertaken previously. From the correspondence between Halley and Burchett, we can follow his progress in these surveys in the summer of 1701 and can appreciate the effort that was expended on this work. But in mid-September he could report, "I have discovered, beyond my expectation, the generall rule of the Tides in the Channell; and in many things corrected the charts thereof."[64]

The result of this survey was a map that seems to have escaped the attention of a number of students of Halley's work.[65] It is called "A New and Correct Chart of the Channel between England & France with considerable Improvements not extant in any Draughts hitherto Publish'd shewing the sands, shoals, depths of Water and Anchorage, with ye flowing of the Tydes, and the setting of the Current; as observed by the Learned Dr. Halley" (fig. 4). This undated chart, probably issued in 1702, is in two sheets each 25 by 19 inches. Again the publisher was Mount and Page but no individual engraver's name appears on the map.

[62] MacPike, *Correspondence*, p. 118.

[63] Ibid., p. 215.

[64] Ibid., p. 120.

[65] Chapman, "Edmond Halley as Physical Geographer," omits reference to the "Tidal Chart of the English Channell." MacPike does not include this item in his list of Halley's publications, as he does other important maps; however, there are some references to it in the writings of others contained in MacPike's *Correspondence*, pp. 9, 22. Hinks's review of Chapman (*Geographical Journal*) deals briefly with it, and it is the subject of a short article by Professor J. Proudman, "Halley's Tidal Chart," *Geographical Journal* 100 (1942), 174–176. Some later writers, such as Bullard and Armitage, discuss Halley's Tidal Chart, and the latter reproduces a portion of it; see Armitage, *Edmond Halley*, p. 149. Since the initial appearance of my Clark Library Seminar paper, part of the Tidal Chart with a description has been published in Derek House and Michael Sanderson, *The Sea Chart: An Historical Survey based on the Collections in The National Maritime Museum* (Newton Abbot, 1973), pp. 80–81.

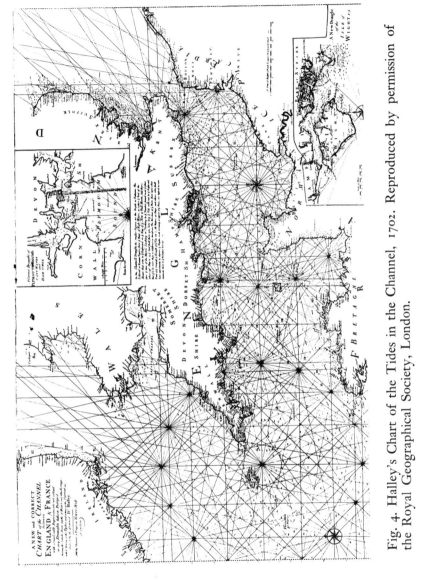

Fig. 4. Halley's Chart of the Tides in the Channel, 1702. Reproduced by permission of the Royal Geographical Society, London.

There seem to be other issues, and it was reengraved for the 1723 edition of *Great Britain's Coasting Pilot* and used in that work for several decades thereafter.[66]

Some features of the chart employ techniques that were cartographic conventions by Halley's time, which he had used before. Thus there is a very liberal scattering of depth values in fathoms, and the map is crisscrossed with lines emanating from over twenty centers. These directional indicators with the two wind roses give the map the appearance of a portolan chart. Latitude is expressed with one degree ticks on the right and left margins of the map considered as a whole. Longitude is not indicated except by some irregularly spaced vertical lines, the value of which are not stated. Neither is the projection noted, but, in any case, projection becomes progressively less significant the smaller the area covered. Careful analysis of the projection indicates no increase of the spacing of the parallels in higher latitudes, over the admittedly limited area shown. Therefore, the Mercator was not used for this chart, and, in fact, it is likely that a plane chart projection was intended. The coastlines are well drawn, with shoals and other features shown. In a letter to Sir Robert Southwell, Halley describes his method of marine surveying in some detail; he indicates his preference for taking angles by the sun rather than the compass and explains his resection method of coastal survey which can be accomplished on a ship under sail.[67] On the tidal charts two insets are included, one of Plymouth, with some verbal instructions to navigators, and one of the Isle of Wight and adjacent parts of the English Coast. There is a graphical scale with twenty divisions of English and French leagues for the general chart; the representative fraction is about 1:1,150,000.

The main theme of the map, which sets it apart from all previous charts, is the indication of the tides. Halley explains his symbolism in these terms on the map itself:

The Litteral or Roman Figures shew yᵉ Hour of High-Water, or rather yᵉ End of the Stream that setts to yᵉ Eastward, on yᵉ Day of yᵉ New & Full Moon. Add therefore yᵉ time of the Moons Southing or Northing to yᵉ Number found near yᵉ place where yoʳ. Ship is, & yᵉ Sum shall show you how long yᵉ Tide will run to yᵉ Eastward. But if it be more than 12 subtract 12 therefrom. The Direction of yᵉ Darts shew upon what Point of yᵉ Compass yᵉ Strength of yᵉ Tide sets.

[66] Robinson, *Marine Cartography in Britain*, pp. 42–43.
[67] MacPike, *Correspondence*, pp. 120–122.

More than thirty arrows showing direction and over fifty Roman numerals indicating time are spread over the Channel from the coast of Kent to Cornwall. Halley's information is in approximate agreement with present-day measurements of tidal phenomenon in this area. To make such measurements it is necessary to anchor; we know from his letters that Halley did so many times. He implies that the end of the tidal flood (stream) is coincident with high water, which modern hydrographers do not accept. Also, Halley's note on the map between Winchelsea and Barque [Berck], "Here the two tides meet," locates this variable phenomenon too precisely. Nevertheless, Halley's tidal chart is a cartographic achievement of great originality and utility.

Halley gave up his active naval command after the Channel survey at the age of forty-six. His interests were focused more particularly on other matters, but he continued to concern himself with cartography and navigation to the end of his life. From late in 1702 to late in 1703 he was engaged in diplomatic missions on behalf of Queen Anne to various European courts, including Vienna. Early the next year he was appointed Savilian Professor of Geometry at Oxford in spite of the testimony of Flamsteed that Halley "now talks, swears, and drinks brandy like a sea-captain."[68] The house with its observatory, which Halley occupied at Oxford, can still be seen.

It was during this period of his life, in 1705, that he contributed his greatest paper on cometary astronomy to the *Philosophical Transactions*.[69] Here we can say no more about this brilliant paper, which was based on the Newtonian planetary physics, except that he correctly predicted that a comet he had seen in 1682 would return to view every seventy-six years. It is through this prediction, almost alone, that Halley's name has become well known. He was awarded the degree of Doctor of Civil Law in 1710 at Oxford, where he continued his major work on mathematics and astronomy, visiting London frequently. It was in the capital that Halley observed the great eclipse of the sun of 1715. He had predicted its track across southern England and published a broadside to inform the public of the event. We may use Halley's own words to explain his interest in this phenomenon:

it has so happened that since the *20th of March, Anno Christi* 1140,

68 Armitage, *Edmond Halley*, p. 156.

69 Edmond Halley, "Astronomiae Cometicae Synopsis," *Philosophical Transactions*, no. 297 (1705), pp. 1882–1899.

I cannot find that there has been such a thing as a total Eclipse of the Sun seen at *London*.

The Novelty of the thing being likely to excite a general Curiosity, and having found, by comparing what had been formerly observed of Solar Eclipses, that the whole Shadow would fall upon *England*, I thought it a very proper Opportunity to get the Dimensions of the Shade ascertained by Observation; and accordingly I caused a small map of *England*, describing the Track and Bounds thereof, to be dispersed all over the Kingdom. . . .[70]

Halley's cartographic understanding shows up well in his map (fig. 5), entitled "A Description of the Passage of the Shadow of the Moon over England in the Total Eclipse of the Sun on the 22nd Day of April 1715 in the Morning." It was engraved by John Senex. The base of this map is the most conventional of all of Halley's cartographic works with its graticule indications along the edges, its neat lines, graphical scale in miles, and well-defined coastline. But again the purpose of the map and its theme are interesting and original. It not only shows the path and shadow of the eclipse, but also indicates the time taken in its passage across England. Perhaps better than any other of his maps it expresses Halley's interest in the entire cosmos, both the heavens and the earth. Since the map was made before the event it depicts, it demonstrates the highest attribute of science—the ability to predict.

Late in 1719 Flamsteed died, and Halley, who was the logical choice, was appointed as the second Astronomer Royal on February 9, 1720. Halley retained his Oxford professorship but now embarked on an ambitious project to observe the moon through the eighteen-year period of the saros.[71] While engaged on this work he lent support to others who were contributing to cartography and navigation. For example, he provided a discussion of projections in *The Atlas Maritimus and Commercialis* of 1728. This work, which was edited by Daniel Defoe, contains a recommendation by Halley on the use of charts for navigators, es-

[70] Edmond Halley, "Observations of the late Total Eclipse of the Sun on the 22nd April last past, made before the Royal Society at their House in Crane Court in Fleet Street London. With an account of what has been communicated from abroad concerning the same," *Philosophical Transactions*, no. 343 (1715), pp. 245–262.

[71] "A regular cycle in the 'sarotic' [saronic] period of 223 lunations (18 years, 11 days) after which the moon returned to approximately its initial position relative to the sun, to its node on the ecliptic and to is apogee." Armitage, *Edmond Halley*, pp. 47–48.

Fig. 5. Halley's Map of the Shadow of the Moon over England, 1715.

pecially for those who wish to engage in great-circle sailing. A few years later, in 1733, he endorsed Henry Popple's map of North America, referring to it as the most accurate map of the area at the time. Halley, like most scientists, felt that the problem of determining longitude at sea would be solved by astronomical means and worked toward this type of solution. Nevertheless, he supported the work of John Harrison on the development of chronometers, which eventually performed within specified limits. Harrison eventually won the great prize offered by the British government for the solution of the longitude problem.[72]

In 1729 Queen Caroline of Ansbach, consort of George II, visited the Royal Observatory. She learned that Halley had been a captain in the Royal Navy and obtained a warrant enabling him to draw half pay for the rest of his life. Halley, of whom it can truly be said that he "warmed both hands before the fire of life," died on January 14, 1742, an octogenarian. He remained a captain in the Royal Navy, Savilian Professor of Geometry at Oxford, and Astronomer Royal until his death.

Of Eratosthenes, another famous astronomer-cartographer of two thousand years earlier, it has been said that he was known as beta to his contemporaries because he ranked second in all his varied professional activities. Later scholars have a higher opinion of Eratosthenes' accomplishments, especially of his measurement of the circumference of the earth. As Walter Hyde has suggested, "At least in the field of Geography he [Eratosthenes] is worthy of *alpha*."[73] Similarly, if Halley was the second of the English natural philosophers of his time, he surely deserves an alpha in cartography.

---

[72] Thrower, "The Discovery of the Longitude," p. 380; however, Newton put the chronometer first among several methods of solving the problem of finding longitude at sea which he discussed.

[73] Walter W. Hyde, *Ancient Greek Mariners* (New York, 1947), p. 14n.

INDEX

References to illustrations are italicized; references to notes (n) are informational, not bibliographic.